S

Rachel Lynch grew up in Cumbria and the lakes and fells are never far away from her. London pulled her away to teach History and marry an Army Officer, whom she followed around the globe for thirteen years. A change of career after children led to personal training and sports therapy, but writing was always the overwhelming force driving the future. The human capacity for compassion as well as its descent into the brutal and murky world of crime are fundamental to her work.

Also by Rachel Lynch

Helen Scott Royal Military Police Thrillers

Detective Kelly Porter

RACHEL LYNCH

SILENT BONES

CANELO CRIME

First published in the United Kingdom in 2023 by

Canelo
Unit 9, 5th Floor
Cargo Works, 1–2 Hatfields
London SE1 9PG
United Kingdom

A CIP catalogue record for this book is available from the British Library.

Print ISBN 978 1 80032 728 3
Ebook ISBN 978 1 80032 108 3

This book is a work of fiction. Names, characters, businesses, organizations, places and events are either the product of the author's imagination or are used fictitiously. Any resemblance to actual persons, living or dead, events or locales is entirely coincidental.

Cover design by Tom Sanderson

Cover images © Arcangel, Shutterstock

Look for more great books at www.canelo.co

Printed and bound in Great Britain by Clays Ltd, Elcograf S.p.A.

1

Prologue

The joke wasn't funny. Not tonight. His habitual clowning around was no longer a source of camouflage that kept him distant and safe. He felt naked and exposed without it.

It was tight in the small dark space and he struggled to get a full breath. Things had gone too far. The rising panic in his chest told him so.

His head hurt and he sensed a sticky substance around his temples, running into his eyes and his mouth. He tasted his own fear. In the blackness, the petrol-like fluid clung to his skin. It didn't smell the same as the grime from the lake. Clarity, which had taken its time to dawn on him, now gripped his senses and he knew, beyond the doubts denying his reality, that it wasn't sweat either. It was just that his reasonable brain didn't want to admit the truth.

It was swelteringly hot in the confined cavity. He calculated that they'd been driving for some while, over bumpy tracks, round tight bends and up and down hills. They weren't in town, of that at least he was sure. He hadn't heard any other traffic for some time and they weren't constrained by junctions, stops and lights. It was oily black and he was completely sightless. His upper body tightened further and he shouted out, but his throat was raspy and sore from the exertion.

He heard a radio and the volume was turned up, because everybody loved the tune, including him. The Police had re-released 'Every Breath You Take' and he listened as Sting's voice comforted him. Nothing could go bad when Sting sang, it was a love song, or so his mum said. She was in love with the singer and the knowledge made him feel awkward. Women her age shouldn't think like that, or that's what his dad said. His parents' faces caused unwelcome feelings to rise up in his already compressed torso. They hung in the gloom as his body rattled around the boot space. He'd let them down.

If only he'd taken his jacket off, like he'd been asked to, he wouldn't be so hot. And maybe he wouldn't be here right now.

The car hit a large bump and it jarred his back, and he cried out once more.

But they ignored him.

The fur inside his jacket rubbed against his cheek, as if it knew that he needed consolation in that moment. He could get another jacket, he thought despondently as his fingers wriggled around the material where it had been cut. It might take him a whole year to save up for one, but he could do it. He'd take an extra job at the working men's club in town. Legally, he couldn't serve booze until he was eighteen next year, but no one would tell if he pulled the odd pint for his dad's mates.

His mind was made up, and the plan was a welcome distraction as he rolled painfully in the shrouded shadows of the tiny metal chamber. In his wild imagination, it became a tomb. The hopeless claustrophobia, and the deep dark terror, was only relieved by the occasional nod to the outside world, as periodic light penetrated the dimness. Tiny cracks in the metalwork focused his eyes as they became accustomed to the endless, crushing murkiness.

A momentary flash of illumination confirmed what he already suspected. His head cracked against the hard interior and made contact with the rough carpet lining, and he knew that the oily substance sticking to his clothes wasn't sweat or mud, or fuel, but his own blood.

Chapter 1

Kelly Porter backed her car out of her driveway and she headed into Pooley Bridge as its residents slowly woke up to another day. The drive to her office in Penrith was a mere twenty minutes. Some days she fancied the route past the great mansion of Dalemain and through the tiny hamlet of Stainton, but today she decided upon the narrower road to the east of the village, through Sockbridge, by way of some diversion. Her mind needed distraction, and negotiating the early morning tourists in their rented vehicles, heading to the hills, gave her ample time to decompress.

Not that she needed a rest from her daughter, or the domestic goings-on brought about by a blended family. But the demands of an eleven-month-old, as well as Johnny's teenage daughter, were sometimes overwhelming, and space was something that was essential for her sense of humour to survive. She checked herself in the windscreen mirror, gently placing a strand of sun-kissed auburn hair behind her ear. Forty suited her. Some would say she was lucky. Her partner, Johnny, enjoyed the luxury of part-time volunteer work with the mountain rescue teams, and so he was a hands-on dad, available to fill in around her more structured role as head of the serious crime unit for the northern lakes. It wasn't an aspersion on his lifestyle, it was just that his routine was more fluid, and thus suited to the needs of a growing family. He could take shifts around Kelly's more formal arrangements, which required her to be in the office every day. Tension around her job was a burden accompanied by a certain amount of guilt, and they'd split briefly for just that reason. Whether they could ride future storms remained to be seen. For now, they were committed to working on their relationship.

The road was quiet, even for peak season, and she drove over the new bridge and parked in the village centre. Darren's coffee, at the Chestnut House, was a treat she indulged in when she had the time. Pooley Bridge was a tourist village, with holiday lets and inns taking up most of the real estate. The typical offerings for visitors were catered for by outdoor shops, bookshops, and souvenir vendors. The Chestnut House comprised a small supermarket, stocked with local produce and items that relaxed holidaymakers thought they could afford: jams, fudge and toffee made in the Lakes, wrapped in pretty boxes covered in images of the Wainwright treks. Through the back, they had a daily delivery of pastries, and Kelly considered the odd croissant or pain au chocolat a little morning pick-me-up when she'd skipped breakfast, like she had this morning.

The fact that Lizzie had wanted to play hide and seek at three o'clock this morning added to her desperate need for caffeine and sugar, and she'd left Johnny sorting washing, with Lizzie helping by placing pants on her head. Their eleven month old was as oblivious to her parents lack of sleep, as she was to their fluid gender roles when it came to raising a child. It had just happened that way, and Johnny was equally as comfortable changing nappies as he was saving lost souls on the mountain-side. His ego was equipped for being a stay-at-home dad.

He knew Kelly went to work for a break.

She parked in the village and walked to the Chestnut House, which was always open early for business. The steamers delivered thousands of day trippers throughout the summer, pouring seasonal pounds into the local economy, keeping the residents afloat, even with the bite of rising costs. Kelly did her bit. This morning she felt like buying two pastries.

'Morning, Kelly!'

Darren was always chirpy. His face never failed to make her smile, which was much needed in her line of work. She enjoyed

4

speaking to real people who were still alive occasionally; it threw light and shade over her existence. Shelley was busy stacking shelves, and shouting questions at Darren from the back. It was a normal weekday morning, despite being peak summer season.

'It's going to be a glorious day!' Shelley said, poking her head over a box of toffee.

'How will that hold up in this heat?' Kelly asked.

The heatwave seemed to have arrived, like a furnace on a dial, two weeks ago, and showed no signs of letting up. The reservoirs of the north of England were already worryingly low, including Thirlmere, to the south, which supplied the city of Manchester. Tourists thought it had always been a natural lake, and not the man-made extension of the tiny Wyburn, carved out by the Victorians and run by a cut-throat business for profit, that it was. Standing on top of Raven Crag, you'd never guess that it wasn't sculpted by a 4-million-year-old glacier like most of the others, because it was so tastefully done, in keeping with the landscape. A pretty stone bridge ran over it, not dissimilar to the new one at Pooley Bridge, and Kelly had great affection for the views to be had from the top overlooking the lake.

At eight in the morning, the ruthless heat hadn't yet taken a hold of the village, but by midday, Kelly knew she'd be thankful for the cool of the old red stone building of Eden House in Penrith, where she spent most of her time. The old walls were merciless in the winter, but welcome in the summer. She wore casual kit: trainers, a comfortable skirt and T-shirt, donning formal wear but rarely, when she had an important meeting or a press conference.

'I quite like runny toffee,' Darren said happily. Shelley rolled her eyes.

Kelly went to the coffee machine and pressed the button for a cappuccino, and chose two pastries.

'Two today, eh? Big job? Serial killer on the loose? Or are you out catching that crocodile that took those swans last week?'

They all laughed. Apparently, a steamer captain had seen three swans disappear under the surface of Ullswater, never to re-emerge. It had been the subject of local gossip ever since. Chances were it was a giant otter, but the internet said it was a crocodile, and the story had stuck.

'Three cancellations this week,' Darren said. He referred to local holiday lets which had apparently suffered from the news of a reptilian killer in the lake. Locals found it amusing, until it lost them money.

'I'll bag these apples myself,' Shelley chipped in. 'Honestly, Kelly, I'm gonna murder him one of these days.'

It was Darren's turn to roll his eyes. Kelly watched her counting fruit, noisily, as if to hammer home her point.

'Be careful what you tell me, Shelley,' Kelly said. 'You could do a lot of damage with those apples.'

'Yeah, she's my witness! You threatened me!' Darren chuckled.

Kelly paid for her breakfast and went out into the sunshine, making her way back to her car. There was something about sipping hot drinks in warm weather that Kelly enjoyed. Weird, she knew, but strangely satisfying. She popped the cup in the holder and started the engine, pulling out and lowering her sunglasses over her bright green eyes. She couldn't wait to eat and devoured the croissant first, dispersing crumbly flakes all over the passenger seat and down her top. It hit the spot and she'd wait for the other delight until she was stationary on the M6 junction. By the time she reached the office, her hands would be greasy, but her belly full, and unlikely to rumble again until it was time for lunch.

As she drove away from the fells, with the national park behind her, on the approach to Penrith, Kelly felt a sense of contentment with her job. It was something that one could never predict. It had its horrors, for sure, and she'd seen her fair share of mangled bodies and the darker side of humanity, but for the most part, it was entirely unpredictable, and that's what kept her in the Lake District. No day was ever the same.

Chapter 2

Kelly's drive into Penrith was without event. Crime in the northern lakes was sporadic. Her team found themselves between the ebb and flow of the larger cases, which sucked their resources and made them dig deep. The in-between times were opportunities for paperwork and admin: dull in itself, but necessary. Kelly tapped her hand on the steering wheel as she waited to get across the M6. She'd never been a patient person. The news was all about the weather, which wasn't uncommon for the Lake District. In the winter, it was all about clouds and storms, and in the summer, sunshine and windows of perfection for the higher peaks. It was all well and good planning a hike up Scafell Pike, but if it was cloudy at the top then your efforts could be well and truly wasted. Climate change was on everybody's lips, and how the hot spells were unprecedented, but Kelly remembered other stifling summers.

The warm periods did make Johnny's job busier though. Hikers always underestimated the fells. Sheer drops and ragged rocks made for tricky terrain underfoot and one tiny mistake could have fatal consequences. But her business wasn't bodies at the bottom of a mountain from misadventure, it was victims who'd met their end at the hands of a perpetrator with intent. Accidents happened all the time, be it at the wrong end of farm equipment, chasing sheep across a bog or cramping in a freezing lake, but Kelly's job was taking up the cases where humans had caused the misdeed on purpose. Murder was a rare malady, though many contemplated it. She recalled Shelley's tongue in cheek threat earlier, and the moment was bittersweet:

most homicides were at the hands of somebody known to the victim. What tipped somebody over the edge from an idle jibe to cold killer was something that had always puzzled her. Much crime was committed in the heat of passion, but there existed a very distinct line between the two sides of the law.

The flow of traffic into Penrith itself was benign and she parked behind Eden House, anticipating a catch-up with her team. As she approached the front door, trotting up the stone steps, a figure appeared and shoved a mobile phone into her face.

'Detective Porter, what's your reaction to the body found in Thirlmere reservoir this morning?'

Kelly's brain raced and she was stuck between brushing the phone away out of her personal space, and processing what she'd just been asked. A body? Why hadn't she been told? She noticed the lanyard around the woman's neck, which said PRESS, and Kelly pushed her way past, up the steps.

'No comment,' she said.

The woman didn't give up that easily and virtually chased her up the stone stairs.

'Could it be a suicide?' the journalist pressed her.

Kelly ignored her. She tuned out from the sound of the woman's voice and went in through the double doors, thoroughly vexed by the intrusion. She greeted the uniforms at the front desk and a young constable stood up and left her desk, coming into the foyer.

'Questions about a body in Thirlmere?' Kelly asked.

'Press on to it already?' the constable asked, peering through the doors. The woman was still there, and she waved brazenly at them both. They tutted in sync, and turned towards the lifts.

'Came in an hour ago, boss. No details yet. A squad car has gone down there to see what's going on.'

'Do I need to be worried?' Kelly asked.

'No idea, boss. It was reported by an early morning swimmer, we haven't even generated any paperwork yet.'

'So how the hell does she know about it?' Kelly asked, indicating the journalist outside. The constable shrugged. 'Find out who she is, she's obviously been given the inside track.'

Kelly marched to the lift. The problem with a county where sheep outnumbered humans was that everybody knew each other's business, or thought they did. It was unhelpful in her line of work. All she was concerned with was if a crime had been committed. A body turning up in a reservoir could have lots of possible explanations, so before the press went on a spree of wild theories, she needed to establish the facts.

The lift opened on the fourth floor and she entered, peering around to see if her second in command had got hold of any information before the press. They wouldn't have long. A lone journalist armed with misinformation could cause a lot of damage, and muddy the facts. A body in a lake in Cumbria was not necessarily news, though she could see the evening head-lines if they didn't get a handle on it. She could even imagine the involvement of the croc supposedly terrorising Ullswater: maybe it had caught a taxi down to Thirlmere and dragged somebody under the surface?

Kate emerged from her office as Kelly threw her things down on a table in the incident room and walked to the coffee machine. Her second in command was trimmer since starting her affair with the superintendent, Andrew Harris. Older than Kelly by a decade, she didn't resent Kelly's rank, nor did she suffer fools. She dressed casually, like her boss, and nursed a coffee.

'Body? Thirlmere?' Kelly asked, eyebrows raised. 'Morning.'

'Morning, Kelly,' Kate said, smiling. 'That was quick.'

'Inside job? Some PC's little sister?'

The journo had been young, all keen eyes and innocent smile. She'd gotten under Kelly's skin.

'Came in just as I did. No details yet. A squad car has gone to investigate and report back. Apparently, a woman going for a wild dip this morning noticed something near the water's edge and rumour has it that it's a body.'

'If it is, we'll look like idiots sending a lone squad car. I've just been stopped by a journalist outside; chances are she's in her car on the way there now. Fancy a drive?' she asked Kate, who nodded and went to grab her bag, plonking her mug on a desk.

'Morning,' Kelly greeted the others.

Rob, one of her longer-serving detective constables, peered up from his computer. 'Heard about the body in Thirlmere, boss?' he asked.

'Late,' she said. 'It appears to be old news.'

'Morning, boss,' said Dan, a more recent addition to her team. He was a no-nonsense Scot from Glasgow who asked the right questions, and didn't shy away from controversy. 'Heard about the body in Thirlmere, boss?' he asked.

'You crack me up,' she replied. Both wore suits without ties, and both towered above her. They'd be her first choice in a sticky spot.

'So, does anyone know anything more than the young Rupert Murdoch downstairs? I've just had a phone pushed in my face.'

'The squad car has arrived, boss. Morning, by the way.' DC Emma Hide appeared, bright and keen, in sports kit, always ready for a twenty-mile run across the fells. 'There's something suspicious at the water's edge, where it's receded beyond record levels,' she said. 'Could be years old. And it's not a body. It's bones.'

Chapter 3

Kelly drove and Kate sat in the passenger seat, keeping her updated from incoming messages. It was a short drive to Thirlmere and Kelly remembered her time in central London with mixed emotions. She certainly hadn't racked up the miles in her car like she did here in Cumbria. Six years ago, when she'd returned to her home county after a stint in the Met, life had slowed down. Bastards doing horrible things hadn't changed: they remained the same, the scenery had just got prettier. But the pace of investigation up here was different, and she spent a lot of time driving. Not that she minded. An excuse to get out of the office and on the road, journeying through the seasons, past fells and dramatic rock formations, plunging down to dark blue lakes, was something that no job spec could advertise. She felt lucky. Mountains had replaced skyscrapers and high-rises, and she didn't mind a bit.

'How's Lizzie?' Kate asked.

'Talking my head off. She babbles what sound like whole sentences at me, even in her sleep, and she's so fast, I can't keep up with her disappearing and bashing into cupboards on that wheelie contraption Ted bought for her.'

'That's my dream,' Kate said.

'What?'

'Revenge. As soon as I'm a grandma, I'm buying the most irritating toys, like drum kits, and things that have loud horns.'

'Nice. Did I do something in a past life?' Kelly asked, scrunching up her face. 'Be careful what you wish for, you're too young to be a grandma.'

'How is your dad?' Kate asked.

Ted Wallis was the senior pathologist, and coroner, for the north of England and Kelly's cases kept his mortuary busy. He also happened to be Kelly's biological father. He was revered and respected in his professional field, but he was also Lizzie's grandfather and the father Kelly had never had. Her relationship with the man she'd thought was her father, John Porter, had been distant and frustrating. She was always playing catch-up to get his approval. Sad how kids go to any lengths to prove themselves to the most unworthy of caregivers. It wasn't that John Porter had been a bad father, it was just that he was no good. More to the point, he'd never been available. He'd been a copper and a good one by all accounts, but a lousy father and a lousier husband. Hence her mother's dalliance with the dashing coroner, Ted Wallis.

Everybody keeps secrets, she told herself.

'He's good. I wish he'd retire but there's no chance of that.'

'He'll want to know about this,' Kate said.

'The bones? Let's take it one step at a time, it's probably a deer, dead long ago after falling in to the water, or dragged in there by the Ullswater beast, who knows?'

Kate laughed. 'Mobile crocodile. Makes you think twice about going in,' Kate added.

'Putting you off your daily champion wild swimming?'

'Yeah, good point, last time I went swimming in a freezing lake was about three decades ago. Are you not slightly intrigued though?' Kate asked.

Kelly and Johnny were keen wild swimmers, especially out of season when you could have a pool at the base of a waterfall to yourself for hours, if you had the constitution, and a thin neoprene skin to hand, in the winter. It was invigorating. The incomparable feeling of plunging into ice-cold moving water, controlling one's breathing and calming the mind, taking the first strokes into the middle of a lake… only to be ruined by the thought that you might be taken by a huge croc at any moment. It had crossed her mind, when the sky was quiet and the mind played tricks. But she dismissed it.

'Maybe the swans were diving for fish,' Kelly said.

'Bollocks they were, they never resurfaced.'

Kelly took the A66 and turned off at Threlkeld, down the road which followed St John's Beck. It was a beautiful day. The sky was bright blue and they had the windows down. It wasn't long before Raven's Crag came into view, and the receding waterline of Thirlmere reservoir. They could already see that a small crowd of people had gathered and they'd heard on their radio that the pair in the lone squad car had requested help erecting a boundary. Maybe it wouldn't be just an excuse to get out after all. There must be something to look at. Kelly turned off and headed towards the small farm at the north end of the lake. It was the best access point to the shoreline. With sadness, she noticed how in the drought conditions, the low waterline exposed the uniformity of the lake bed and revealed its true purpose. She drove over the small stone bridge and parked at the other end where two police personnel in uniforms were struggling to prevent members of the general public going any further. They faced various jibes and complaints from visitors today, because this was the best place from which to approach the fells above the reservoir, and it would be shut for at least the whole day, if not longer, depending on what they found.

The small group of people stared at Kelly's vehicle and began to gossip, realising they must be important to be let through the tape put there by officers who weren't answering any questions. She didn't need to show her lanyard. The Lakes crime unit was small and Kelly was well known, thanks to a few recent high-profile cases. A few bystanders pointed at her car. She parked beside a large pile of seemingly discarded farming equipment and they got out. It was dry underfoot and the temperature was rising rapidly. The sun was hot and bright, and Kelly placed sunglasses over her eyes. One confident individual shouted a question but she ignored them, scouring the crowd quickly for any suspicious behaviour. Several phones were held aloft in the air, taking snaps to be posted on social media; their moment of fame.

A uniformed officer lifted the tape for her and Kate, and they were directed down a dusty path towards the lakeside, where another officer stood with his hands on his hips, looking puzzled.

'Ma'am,' he said in greeting.

'What's all the fuss about?' Kelly asked.

She needn't have waited for an answer, though.

As they approached the receding lake bed, Kelly felt as though she were somewhere else, not Cumbria, the green and lush land of forests and waterfalls, but somewhere arid and forbidding. The lake hadn't looked like this for years. She'd heard that reserves were at 50 per cent, and now, peering at the sad pool of water in the centre, she wondered what else they'd find in the coming weeks. Stories from around the world reported dead pool levels, when water volumes were so low that rivers couldn't flow downstream. But the water authority had reassured the general public that they weren't close to that here.

On the ground, close to the water, vaguely resembling a human form, were several bones in a misshapen formation, some sticking up to the sky and others randomly languishing close by. They were a shade of ginger-yellow, and reminded Kelly of walks along faraway beaches, where driftwood poked out of the sand, waiting to be taken home and planted by loving gardeners, as a feature. Kate walked around the remains and Kelly knelt down to peer closer. She was no anthropologist, but she knew a human skull when she saw one. She identified the jawbone straight away, and took off her sunglasses to examine the specimen, without touching anything, knowing that they'd have to get some experts down here. The teeth were still attached, yet wonky and crowded, in what would once have been the gum, and she noticed several old-style amalgam fillings. Ribs stuck straight out of the dried muddy bed, and further down, she could see a hip bone and a further set of bones attached to it. It was incomplete, for sure, though the site would have to be excavated properly to tell. There were odd bits of detritus laying around and she noticed pieces of yellow cloth.

'Drunken accident?' Kate whispered. 'I can't remember any significant missing person cases in the surrounding area in the last ten years,' she added.

Kelly nodded. 'We'll have to check. There's not much for us to do except call in forensics and the coroner, though I'm sure he'll want to get a forensic anthropologist down here. Looks like we'll have to seal off the area for quite some time,' she said. They stood up.

'It's so tempting to dig, isn't it?' Kate said.

Kelly agreed. 'Especially that arm,' she said. The skeleton was arranged well enough for them to make out the human form, and though missing a leg, and quite a few ribs, the arms looked intact, though in an unnatural position. Whoever it turned out to be was probably still relatively together because there were no predators in Cumbria's lakes, though the Ullswater croc might have had a good meal off it. She knew that roughly speaking, human bodies skeletonise within around two years, but in cold water, who knew?

They looked closer.

'Are you thinking what I am?' Kate asked.

'It's an odd position. I'm no expert,' Kelly said, 'but it looks to me that the arms are both behind the body. Give me a minute.'

Kelly paced away and called Ted Wallis.

'Dad, you've heard?'

Kate listened to the call; it was clear that Ted, like most of Cumbria, had woken up to the news of a skeleton in Thirlmere, despite the fact that Kelly had only just got down here. Kate watched as Kelly nodded and went back to the remains, bending down once more.

'Yes. The arms seem to be behind the back in an awkward position, but they disappear into the dry mud bed, so I can't see much,' Kelly said into her phone.

She moved closer.

'Yep, the teeth are intact, so we should get a dental ID. Hold on. No, no clothes as such, as far as I can see, but we do have remnants of material here and the surface of the bones look soft, as if something is glued to them. Could it be clothes?'

She looked at Kate and nodded.

'Kelly,' Kate said quietly. 'Look.'

Kelly squinted to where Kate was pointing and she saw it. It was easily missed. The metal was covered in the same grey-brown dust as everything else, but the uniform shape was distinctive.

'Dad, I think we have a bracelet.'

Chapter 4

Ted Wallis had seen plenty of skeleton remains in his forty-odd years as a pathologist. This one intrigued him. Not only was it his daughter who was waiting for him beside the body, ready to be examined, but it was already a mystery. He approached the quandary like a scientist, with a head full of questions to eliminate variables. Only that way could they ascertain who it was and how they got there. Theory wasn't something Ted accepted. He needed hard facts, and that would take time. It would help, of course, once they had a dental ID. That would at least give them a timeline.

He normally wore a suit and tie but the weather was insufferable and he'd finally conceded that in this heat, his old body moved better in loose clothing. He wore chinos and a grey shirt, open at the collar, and based on what Kelly had told him about the lake bed, he'd opted to wear a pair of trainers that he only wore when he walked in the hills. Boots were uncomfortable and if Alfred Wainwright could scale the peaks in tweed and a flat cap, then he could also wear what he damned well like. He felt distinctly unprofessional in the casual wear, but field work called for appropriate attire.

By the time he arrived at the dry lakeside, the crowd had grown bigger to the north-east side of the great dam, which had been built by the Manchester Corporation in the 1880s. He noticed members of the press queuing up and angling for a photo opportunity, but they were being kept on the east side of the dam. There were laws directing the UK press not to show bodies, and so they wouldn't be able to print much, even if they got close. However, they were interviewing anyone willing to give a soundbite, potentially hampering his work. But he was used to blanking out all background noise. He peered up at the fells above Thirlmere and noticed that somehow, walkers had managed to negotiate the forest behind and were lining the walkways. It took a while to get through the throng of people and press, but once he explained who he was and who he was meeting, he was allowed to proceed and parked next to Kelly's car. He'd been told that the body lay on the west shore of the reservoir.

It was a different scene than the one Kelly had described earlier.

A forensic van was there, a dozen uniformed police, a proper cordon, and several blue tents, with people pacing up and down to and from the lake in white, full-body overalls.

Despite his own personal discomfort due to the weather, it was a fabulous day for walking, and to the east, he was sure that Helvellyn would be busy. Thirlmere was, in his view, the best starting point for the mountain, and he could see dots of colour denoting walkers' bright clothes above the treeline, probably blissfully unaware of the drama unfolding below.

He spotted Kelly and Kate and greeted them both.

Kelly led him underneath a tarpaulin entrance to a blue tent and they went inside. He instantly felt cooler, but it was also stifling and muggy inside the plastic tomb. Lights had been erected and powered by a generator hooked up to the farm.

'Forensics are doing a sweep of the surrounding area, and they've found some items of clothing already. They might not be linked to this character, obviously.'

'Have you given him or her a name yet?' he asked.

'I've heard Captain Jack Marrow mentioned,' she said, pulling a face.

'Because of his missing leg?' Ted asked, pointing at the lower body. Kelly shrugged. 'I thought as much. Well, I can tell from the pelvis that it's a man. Females have a larger sciatic notch, and he's got a V-shaped subpubic angle. Though he's not fully developed.'

Kelly's stomach turned over. 'A child?'

'No, a teenager perhaps, or a small male. It's difficult to tell because diet and environment can affect these things. You said there was a bracelet?'

'Here,' she pointed to the man's wrist. 'I assumed it was a female,' she added.

'Understandable.' He bent down.

'I've been informed by Eden House records that we have seven significant missing person cases dating back over the last ten years.'

'You might well have to go back a bit further than that. In this freezing deep water, it would have been weighted down and decomposition would have been slow. I see what you mean about the arms. A body which fell into a lake, or was put there, might have moved around considerably with the water and the effect of decomposition gasses. It's not a natural position to settle in. I've called an excavation team from the University of Lancaster, they should be on their way. The remains can't be moved until they've done their job properly.'

'I thought as much. How long do you think until we can get it to a lab for ID?'

'If they start this afternoon, and it's straightforward, then you should have some answers later today, or tomorrow. I wouldn't hold your breath. Those missing person cases, though, anything interesting?'

'It could be any of the males, I guess, if you're sure. I'll get them to go back further.'

'I'd do that. I'm 99 per cent sure it's male. Remember the body in Wastwater a couple of years back? That'd been in there for twenty-five years – you'd be amazed how good old bones look,' he said. 'The space out there isn't big enough for what we need down here,' he said, nodding to the small yard beside the farm. 'You're going to have to clear all the way up to the road.'

She agreed. 'I'll get on it. Are you staying here?' she asked.

'Absolutely. I want to meet this guy, whoever he was.'

Ted put on some plastic gloves and bent over the body, feeling gently around the sandy deposits clinging to the bones. He rubbed the metal of the bracelet gently, revealing silver links. It seemed to hang around the upper portion of the lower arm, and Ted removed it carefully.

'It might have some kind of personal engraving on it, I'll bag it.'

'What do you think to the chances of determining cause of death?' Kelly asked.

'That'll depend on the state of the skeleton. Any natural causes will have to be ruled out, having rotted or been washed away long ago. The forensic anthropologist will be looking for any unusual marks on the surface of the bone, once we get him out of the mud. I've chosen someone who worked on a similar case in Norfolk some years back. A man went missing seventeen years ago and ended up at the bottom of a canal, only to be found when it was drained to dredge for another body in a recent case. It's amazing what you find when Mother Nature decides it's time. I've notified an osteologist too. If these bones are ancient then it'll be over to the National Trust to bring in archaeologists. That's interesting,' he added, staring at the remains.

'What?' Kelly asked.

'There's a fair amount of adipocere.'

'What's that?'

'We call it mortuary wax. It's essentially the leftover fat of a body. Well, it's not left over, it's a permanent cast of the stuff that remains from the original fat. In water, it doesn't putrefy like the other tissues, it anaerobically transforms into a hydrolysed chalk, like soap. It takes years. Look.'

Kelly went towards him and peered at the body. Ted pointed at lumps of a wax-like substance that adhered to some of the bones.

'And what causes that?' she asked. The gloop was what she'd first guessed was the remains of clothes sticking to the bones.

'Saponification. There's not much, but that fits with this being a young male. It's what happens to a body when it's not exposed to air and normal putrefaction.'

He took a small tool from his pocket and began scraping cautiously around the lower arm bones.

'Does the colour determine how long it's been in there?' Kelly asked.

The candle-like substance was grey and the bones a shade of ginger, and it made them look like tan leather.

'Not particularly,' he said.

'The photographer is here,' Kate said from behind them.

A small young man entered the tent and introduced himself. He wore glasses perched on his nose, which kept slipping off in the heat, and he had the air of a university nerd. He glanced at the bones and Kelly thought he might rush back out again. His behaviour gave Kelly the impression that this was his first forensic job. He was armed with several cameras and Ted set to work directing him, and showing him what he wanted photographed, and in what order. It was vital that the position of the body was recorded, so by the time they reached the lab, the exact site orientation was known. The young man also carried a scanner, which would provide a three-dimensional image of how the body was lying, even under the mud.

'There's a fair amount of disarticulation indicating that the body has been here a while,' Ted spoke softly and Kelly watched him as he instructed the young photographer, who'd composed himself quickly, to the correct angles.

Ted leant over once more and groaned. 'My bones feel older than these laying here,' he said. 'Be careful where you stand, there could be dozens of tiny bones underneath the site that have become disarticulated,' he instructed the young photographer.

The young man carefully tiptoed around and stood exactly where Ted said. Kelly stood back.

It was painstaking work. As the camera clicked, Ted scraped the first layers away around the arms, a millimetre at a time, and, after a good half hour, he stood up and exhaled.

'There's your first answer,' he said, stretching his back.

Kelly walked around the body and looked to where he pointed. He'd scraped enough sediment away from behind the body so she could see that whoever this man was in life, he'd gone into the water with his hands tied behind his back.

Chapter 5

Jason Cooper nursed a pint between his hands, and sat in his usual spot: propped up on a bar stool at the polished wooden counter of his local watering hole, like he had done for the past twenty-odd years. If somebody had asked him when he was seventeen years old how his life would go, then sat here, by the empty hearth of The Swan Inn, was perhaps the last answer he'd have given. But that was a lifetime ago, and time could be a punishing mistress when you grow up angry.

Today he was even tenser.

His knuckles were gnarly and his nails thick with dirt from the farm. He'd laboured there for the past four years. Turning forty, and looking ten years older yet, was about as depressing as he thought life could get. It was only now that he was beginning to realise what his teachers banged on about in school, and why they warned him that he'd never make anything of himself. A broken marriage, two kids he never saw, too much drink and zero preparation for the financial burdens of later life, all conspired to make him keenly experience the weight of failure.

He drained his drink and pushed the empty glass towards the young woman behind the bar. She said nothing and took it, filling it with local bitter, skimming off the top, just how he liked it. Booze was the only thing that made him feel anything approaching human these days. It also made the demons go away.

'You going to college then, Connie?' he asked the girl.

'Nah, it's boring. I'm gonna be a hotel manager.' She chewed gum and he was reminded of a cow mulching cud.

'You should get your education,' he told her.

She rolled her eyes. 'Like you?' she asked impertinently.

He smirked at her temerity. 'Now then, it's because I'm a bum with no qualifications that I'm telling you not to be like me. I'm a walking example of how to screw up your life.'

Connie failed to grasp the gravity of his intentions. He meant it. He'd spent the bulk of his shift today on the phone, trying to make amends for his almighty cock-up. To no avail, as usual.

'I'm not going to *screw up* my life. I've got an interview in Windermere. I'll start in the kitchen and be a manager when I'm ready,' Connie said. Her chin jutted out and Jason was reminded of how he used to puff out his chest, when he knew everything about everything. If only he could go back about twenty years and do it all differently.

'Course you will. What do I know?' He winked.

His concentration was pulled away from his pint to the TV that was mounted on the wall but silent. Some journalist was waving a microphone into the face of a dude in a white plastic suit. He recognised the lake as the Thirlmere reservoir.

'Hey, Connie, turn that up will you?' he asked.

She blindly took the controls from behind the bar and pointed, pressing the volume switch. There was no one else in the pub. He could see that she wasn't interested in the news, just like he wasn't when he was her age. The only things that counted were vinyl, cigarettes and what was in between a girl's legs.

He listened intently and time froze. He knew what the piece was about. It was all anybody had talked about all day. The body in Thirlmere. The footage of the lake became one giant canvas in his head and he quickly zoned out of his surroundings. The water's edge, different but the same. Raven Crag, high up on the western fell, the trees and the stone bridge. He licked his lips and realised that his pint had spilled a little. He took a good gulp and downed a half pint in one. He put it down on the bar and released his hand. A warm but not welcoming heat spread across his gut and he thought he might vomit.

Hold your nerve, he told himself, echoing what he'd been ordered to do all day, over the phone, despite telling them it was all over.

Then the cameras focused on Kelly Porter.

God she looked good. That girl had looked after herself, and she'd paid attention in school. Everybody called her Little Miss Perfect. Her old man was a copper and she was all A grades and ambition. Word had it she'd pissed off to London, but he knew she was back because she'd been all over the news in the winter. She'd dealt with that helicopter crash up on Scafell Pike. She was what was known as a dark horse, all right.

If only when you were a kid, falling asleep in a maths lesson, listening to the dumb teachers droning on about shit that made no sense, they could fast forward and show you how you'd turn out. Now that would wake kids up. Kelly Porter had that secret ingredient. She'd always had drive. Always had that Hollywood smile. She turned him down once and he'd never asked again. He was beneath her, he knew that. She was more suited to posh twats from London, with their fancy cars and penthouse flats, no doubt. She wasn't interested in local folk. That's why she'd dumped Dave. So why had she come back? He'd heard her mother died not long after John. John had been a local boy, just like his own father. Loyal, traditional and suspicious of city folk. He was a good bloke. John had got him out of some scrapes.

He watched the TV, mesmerised by Kelly, and remembered her how she was when they were seventeen. She was a loner. Not quite ignored, but never fitting in either. If there was a magic dust that some kids get given by the almighty at birth, then Kelly had it. Everybody knew she'd be something. Somebody. And now she was on TV, smart as you like, holding her own.

He felt something akin to pride and wondered what John Porter had thought of his daughter. Shame what happened to him. Cancer is a bastard. It got Wendy too, he heard. Maybe it runs in families. He wondered what was worse: fuckwittery or cancer being passed down the generations, and decided he couldn't tell.

He needed a piss and slid off his stool and walked past the TV. As he passed under the flat screen, he stopped dead and looked up when they mentioned human remains in the reservoir.

'…it's unclear at this time where the remains are from, or how old they are, and even how they got there. Detective Kelly Porter has told us that they're not ruling anything out until a qualified forensic anthropologist has examined the body…'

Jason held out his hand to steady himself and realised that the last pint he'd ordered was his fourth and he was well on his way to being inebriated. His bladder willed him forward and propelled him into the bathroom, just in time for him to churn up his lunch into the urinal. It stank of beer and eggs. He took a long and satisfying piss, trying to wash away the sick with the flow, then zipped up his fly and went out the back door. He could settle up later, it wasn't as if he was going anywhere.

He walked to his car and started the engine, pulling out onto the tiny single lane road that led to the A66 towards Keswick. His wits were still sharp and he'd driven the stretch of road a thousand times before, but his mind wandered to the Thirlmere reservoir, and the face of Kelly Porter.

She'd done homework for him and helped him in class.

It was a curious memory, and one that he'd long blotted out. After all these years, such a simple thing resurfacing surprised him. He shook his head and tried to concentrate on the road. He knew he was way over the legal limit and he slowed down. He passed a lone police car and his heart jumped into his mouth but it continued on its journey, disinterested in him. He swerved again as he reached for his phone, bashing in numbers known to him by heart. But they no longer answered their phones. He'd spent hours calling them, behind sheep dips, from the wheel of a tractor, in the middle of a lavender field, only to be told the same thing, over and over.

Finally he reached the outskirts of Keswick and pulled into the holiday park. He lived on the edge of the forest, away from the tourists who found the small cabins cute and cosy for a week. That was the deal: he stayed out of the way and didn't bring the reputation of the park down. The owner, Michelle, was doing him a favour. The least he could do was follow her rules. She'd always looked out for him during his darkest moments, and she was there to pick up the pieces when his life fell apart, as it invariably did from time to time. He lived in the tiny cabin rent free.

By the time he parked his car round the back of his temporary accommodation, his head was sweaty and his heart was racing. He got out of the car and slammed the door shut, eager to get inside and flick on the TV. He knew he had a few beers in the fridge and there was a half-bottle of white rum he'd sneaked from the pub one night, some wealthy holidaymaker having ordered it and left it on the table as if to flaunt the fact that they could.

He looked around, the sun casting shadows around him as it moved behind rare clouds.

Michelle had given him the furthest trailer from the reception area that she could and it did him a favour, he kept himself to himself. He was glad to walk around the back, to the door facing the woods, and disappear into oblivion for the evening, until the torture started all over again tomorrow. The only reason he knew he was still alive was the stink of sheep shit on his T-shirt from the farm.

He didn't make it to the door.

The spade hit him fully on the back of the head and he fell forward, bashing his face on the small steps leading up to his door. He was completely unaware of the edge of the tool slicing into the back of his neck, finishing him off for good measure.

The blackness was complete.

Chapter 6

Ted Wallis shook hands with his friend of more than four decades, an esteemed forensic anthropologist who'd shared no less than three flats in London with him when they were struggling students, scraping money together for pints and sausages at the cheapest watering holes in Whitechapel. In those days, the area was still considered to be one of the red-light districts and gangland turfs of London. How times had changed. Now, east London had been overhauled, and one of their old blocks of flats razed to the ground. Jack the Ripper tours, fancy delis and clean streets had wiped away the more unsavoury history of the place.

They were the good old days.

'Not retired yet, then?' they asked each other, and chuckled.

Ted had waited by the lakeside of Thirlmere in anticipation of the arrival of Dr Henry Dempsy, and now it was as if they'd seen each other for a pint just yesterday. They had a lot of catching up to do, but that could wait. Henry was of the same cloth as Ted, and he wanted to get straight to work.

It was imperative that the site was preserved intact around the body as much as possible. The real excavation would be done in the lab, back in the mortuary at the Penrith and Lakes hospital. Henry ducked his head under the tarpaulin and Ted showed him into the tent.

'Nice excuse to come to the Lakes,' he said. 'I've been meaning to pop up here ever since I heard you were based here.'

'That's a long time, Henry,' Ted said.

Henry raised his eyebrows.

'Lovely part of the world.'

'It is indeed. The inside of a tent is not really the tour I had in mind for you though. When this is over, perhaps I can take you for a hike and a pint.'

'I'm not sure these old bones are up to it. Talking of bones, what have we here?'

Ted approached the team of excavators from Lancaster University, who'd arrived in a minibus just hours before. They were like kids on a field trip, excited by the opportunity to unearth a real skeleton, and not just look at one in books.

'They're doing a good job. We retrieved a silver bracelet, which has been sent for analysis, but initial observation places it in the modern world at least. It has a hallmark on it which is consistent with a factory outside of Birmingham that began production in 1979.'

'From the protruding brow ridge I'd say it's a young male,' Henry said.

'That's what I thought.'

'Hands tied behind the back. Homicide, then?'

Ted nodded. 'One doesn't go for a swim without the use of the limbs.'

'And not a suicide either. Local police any good?'

'My daughter is the lead detective and we've worked together for six years now. She's the best, though I am slightly biased of course.'

'I'd like to meet her. Mary must be proud,' Henry said.

Ted inhaled a breath and decided that it was too complicated to explain that Kelly wasn't the daughter of his wife, Mary, but the product of a fling with the woman he'd been in love with for forty agonising years, before her untimely death almost three years ago. That, along with their pint, could wait.

Henry popped on some blue latex gloves from his pocket and moved closer to the remains.

'I had a case like this last year. The body had been submerged in cold lake water, in Canada, for sixteen years. The bone structure looks similar, but of course I won't know for sure until I get them under a microscope. There's cloth adhered to them, which is normal, and the adipocere confirms it was underwater. However, if it was simply floating around and sank, I'd expect none. It's more common when the body has been buried under the water table. Have you taken samples?'

Ted nodded.

Henry bent over the body and touched the earth surrounding it. Three students had been asked to stand back to make room and he examined their work. He felt the mud and tapped it as if he were listening for wall cavities inside a loft. He peered closer.

'Sir, it looks like the soil around the body has been disturbed, it's not as compact as the mud over there,' one student said.

'That's absolutely what I'm thinking,' Henry said. 'It looks distinctly like a grave, but that's impossible at the bottom of a lake. There might be some other explanation.' He turned back to Ted. Henry gazed at the skeleton intensely, and Ted thought how odd they must look, when faced with the dead, as if admiring them or waiting for them to speak.

'I have to say, I'm astounded by how well it is preserved,' Ted said.

'It's the cold,' Henry said. 'This water is normally deep, saline free and gently flowing from up on those hills, it's perfect for preservation, and it's probably what weighed the body down, which is why it's stayed pretty much together. Believe me, this is a fine specimen,' Henry said. 'That broken rib could be trauma from heavy water crushing it against a rock, or it could be ante-mortem and intentional. I'll have to soak the adipocere off to get to the bones properly,' he added.

Ted nodded again, satisfied with himself that he hadn't made any glaring blunders while examining the skeleton himself.

'It's timely, I'll give you that. There's groundbreaking research into deamidation – that is the chemical change in bone proteins after a given time underwater – post decomposition. It can pinpoint the time of death. In clear lake water like this, even though it's a reservoir, the deterioration of bone protein is quite specific, you'll be glad to know. What is the material, do you think?'

Henry pointed to the restraint around the upper limbs of the victim. It had the appearance of fine slivers of dark brown layers of pastry, or at least that was Ted's first impression, but then he had a habit of comparing items to food, especially French patisserie.

'It looks like leather.'

'Well, it's possible. Before mercury tanning was phased out in the 1950s, plenty of the leather in circulation was bomb proof. Nowadays, the plastic cable tie is the restraint of choice for any decent serial killer. However, it could be ritualistic. Absence of chain, as in this case, given that locking somebody's hands behind their back is so much easier than tying them with a bit of leather – and getting it tight enough – is associated with a personal motive.'

'Fits with the bracelet, post sixties at least,' said Ted.

'So that should help with carbon dating if we can't get a DNA match,' Henry said. He stood up straight. 'Excellent specimen, though not whole, this is very exciting.'

Ted appreciated Henry's professional passion, but he also had reservations about the potential tragedy they were unearthing. Henry was right, thanks to the purity of the lakeland water, the bones were still dense. There were no copper deposits, iron ore, or waste products from nearby factories to consider, and so the bones looked in good shape, although that might all change when they tried to move them.

'I'd get the whole lot to the lab – including a foot around the body – and we'll start first thing in the morning. Of course, we can get the mandible off without too much effort. If we're 90 per cent sure this is modern, then you might save us all a lot of time by getting a match with dental records.'

Ted watched as the doctor examined the fragility of the mandible bone and its connection to the rest of the cranium. All

31

cartilage had disappeared, and it was possible that the skull bones had simply been held together by the mud at the bottom of the lake. As the lake dried out and the wind blew it away as dust, the facial bones were revealed in virtually the same position as when they had finally rested on the bottom.

The bone popped off, and Ted logged and deposited it into a plastic evidence bag. It would be sent to a forensic dental lab in Manchester. The lower jaw would be X-rayed and compared to the national database. Amalgam fillings, or the use of a combination of metals to plug decaying teeth, was a practice dating back a couple of hundred years in Europe. Before that they were silver. It was more evidence to suggest that the skeleton was relatively recent. If it turned out to be from the last seventy years then the police would have to formally investigate, and Ted was torn between wanting Kelly not to have to deal with it, and his fascination about who this young man was.

It was out of his hands.

Chapter 7

Captain Jack Marrow was transferred to the mortuary of the Penrith and Lakes hospital, by coroner's van, at ten o'clock in the evening, enclosed in a wooden coffin to make sure that the earth around him was preserved. The flurry of press activity when the large transit vehicle backed up to the tent was sure to make the evening news, at least locally, and possibly the national news, first thing in the morning. Bodies retrieved from lakes made headlines. It set off the macabre and maudlin fascination of the general public and inspired endless conversations about who it could be. Was it an ancient traveller from Asia, weary from the trek across Europe, witness to quite a different Lake District before forest clearance and hikers in polyester? Or was it a more recent missing person? Before the police provided an official statement, theories and wild urban myths would abound.

The air was still warm and the sun was just beginning to dip behind Raven Crag to the west of the shore of Thirlmere. The Manchester water board had risen to the occasion and had imposed solid cordons around their land. Anybody passing the lake, say coming down off Helvellyn, might wonder if there was some kind of high summer pagan festival going on at sunset, down on the shoreline of Thirlmere.

As the van made its way through the clogged lanes, saturated with press cars and journalists, Ted and Henry emerged from the tent, exhausted. Kelly had called Ted seven times, each time informing him of updates from witness statements and asking if he'd discovered anything more of interest. He had little to tell her.

He told her his theory, backed by Dr Dempsy, about the victim being late twentieth century, but warned her that theories were just that: conjecture and supposition.

She'd invited them both over for dinner but it was late by the time they'd left the site. Ted's car led the way and Henry followed behind. As expected, Henry hadn't booked a hotel for the evening, and in his typical style, Ted knew that now he was here, after a gap of possibly twenty years, they would find it difficult to get rid of him. Ted offered his place to sleep and Henry gratefully accepted.

'It'll be like old times,' Henry said before they each climbed into their separate cars.

Ted had a flashback of their shared flat in London, when they were in their twenties; the parties and recreational drugs. The two of them sharing a house nowadays would be closer to a museum piece in appearance and character. The chances of them staying awake past the hour were slim, but Ted had promised Kelly he'd pop in.

The sun setting in the west made for a dramatic drive as Ted headed east, with Henry following, and an orange glow radiated from behind them like a comforting blanket. Ted was chilled to his bones and he ached like the old man he was, despite the mercury not falling below twenty-five degrees.

It was still just light when they pulled into Pooley Bridge and Ted drove past the steamer jetty and pulled into Kelly's driveway.

The door opened and his daughter came out to greet them. She'd been waiting and watching out of a front window and Ted smiled at her. She gave him a hug and he felt a little warmer, just from seeing her. Henry parked beside him and got out. Kelly's driveway was wide, and the properties along this stretch of road roomy, perfect for holidaymakers arriving by car or minibus.

Kelly held out her hand for the man who accompanied her father and told him she was looking forward to hearing his conclusions, though they were slim at this time. She ushered them in and Ted watched her sizing his companion up, her detective eyes never resting.

'Is my granddaughter awake?' Ted asked.

Kelly smiled. 'Always.'

Lizzie had taken to stretching the evening out like a professional. Josie, Johnny's daughter, got up from her position on the floor, where she'd been playing with the child, and went towards Ted to give him a cuddle. She was almost a woman and had lived with Kelly and Johnny for four years, since she was fourteen years old. If she was awarded her predicted grades in August, she'd be off to study history at Durham University and her life would enter a new chapter. It made Ted feel ancient to witness the dawning of a new life. And then there was Lizzie, eleven months old, shuffling around, pushing wheelie toys, and trying to form real sentences. Ted had a nostalgic moment as he cast his eyes over his family, a family which had surprised him when he was thrown into it blindly almost four years ago. Since then, he'd watched it grow, and he felt an urgency to make the most of it before he too passed into the next phase of his life. He watched Josie's ease and grace, and was struck by the confidence of the young, as if they had all the time in the world.

Johnny came downstairs and shook Ted's hand, doing the same to his old colleague. They instantly had things to talk about. With his casual shorts, unshaven face and flip-flops Johnny gave the impression that he was a mountain man, and that was all, but Ted knew that his life too had been pocked with the scars of reality. Johnny offered drinks and Kelly ushered them out to the terrace, where they'd taken blankets out earlier for the dipping sun. Henry was instantly at ease. Ted went to Lizzie and knelt down, holding her outstretched hands as she thrust her hips this way and that. She spent her life upright, whizzing around in her walker, exploring and touching, just like her mother. Or how he imagined she'd been anyway, and Wendy had confirmed it.

'She's still up?' Ted asked.

Kelly smiled. 'She's excited. She's just finished her bedtime bottle and she'll be going to bed anytime, but now you're here, I doubt that.'

Suddenly, from their place of death and unreality this afternoon, they were thrust into a busy and noisy sensory assault, and Ted could see that Henry was just as welcoming of it as he was. Several conversations started at once and the hum of conversation filled the living room, the terrace, and the kitchen. Ted sat on the sofa and Josie got Lizzie out of her walker and plopped her in front of her grandfather, where she held on to him and told him jumbled stories in a language that made no sense whatsoever, but he pretended that they did. Kelly placed a glass of red wine beside him at one point and then disappeared with Henry outside, to talk.

Johnny said that there was food left from dinner, should they be hungry, and Ted realised that he was. He struggled to get Lizzie back into her walker, not only because she was heavy, but because she kicked her legs and didn't want to go in. Josie carried her and sat with him in the kitchen while he ate. Johnny took a bowl outside for Henry.

The weariness of his body began to subside, and he was surprised to look at his watch and see that it was gone midnight. He didn't feel tired. His family had revived him, as they always did.

'How's the new holiday let?' he asked Johnny, when he came back in.

Johnny had purchased a fine townhouse in the middle of Pooley Bridge during the spring, and had been working on making it ready for the first guests later in the summer. Ted had seen it. He was doing a fine job. The summer trade would more than cover the small mortgage he had taken out, and it was an investment.

'It's finished. It'll be ready for our first guests next week, and it's about time because we've missed almost the whole of June.'

'Don't worry, you'll make up for it.'

'Kelly made the spare room up for you, what's your pal doing?' Johnny asked.

'I was going to take him to mine.' He yawned.

'Well, you're both welcome to stay in the new house. You can be my first guests,' Johnny said.

It was a tempting offer, and Ted didn't need to be polite with his almost son-in-law.

'Splendid. That would save me a drive back to Keswick, and I'll get Henry set up at mine tomorrow.'

Johnny nodded and passed him a set of keys. 'I'll walk you round there when you're ready. I think I've remembered everything.'

The house was a five-minute stroll from Kelly's place, and Ted suddenly felt tired again.

'What's he like, your pal?' Johnny asked.

'Who, Henry? We go back years, decades, even. He's a typical professor of the grim arts. He loves inspecting old bones and digging up grave sites. We have a lot in common.'

'Nice,' Johnny said, and smiled. Ted could be open with his proxy son-in-law, who'd served in the army and seen plenty of trauma for himself.

Josie wandered into the kitchen holding Lizzie, and Johnny got up to relieve her.

'Right young lady, it's way past bedtime for you.'

'She had a long nap this afternoon,' Josie offered apologetically.

Though Lizzie was not yet proficient in her native tongue, she most certainly picked up on the word bed, and the change in tone indicating that she was being taken away from her family. She arched her back and cried, and Johnny rolled his eyes. He was a picture of calm though, Ted noticed, and he had a soft touch with his second daughter. He carried her around the guests, saying goodnight, and went out to the terrace to her mother, and Ted and Josie were left alone.

'You know Henry would be a good person to know if you did any ancient history modules. Many primary sources are based in archaeology,' Ted said.

'I know, Granddad. Do you think he'd mind? Have you been with that dead body all day? Do you think it's ancient?'

Ted knew too late that he'd opened a can of worms, but she was eighteen and more than capable of handling the truth. She wasn't stupid.

'Yes, we have. I called him in because he's got osteological experience too. He's worked on burial sites across Persia and South America.'

'Is it a burial site?' Josie's eyes lit up.

'I don't think so. I can't say too much, because Kelly will more than likely take the case, but we think it's twentieth century.'

'So it's one skeleton? My friends were saying that it was a human sacrifice pit dating back a thousand years.'

'I'm sure we'll hear lots of stories like that over the coming days. Don't let on you know the truth, will you?'

'Of course not. So, who is it? Do you know?'

Ted shook his head, finishing his current glass of wine. 'No. Not yet. You know, a lot of what we know about history is from skeletons and burial sites.'

'And poo,' Josie said.

Ted laughed. It was true. Petrified excrement was pivotal in finding out what people ate, how they lived, their environment and even their cause of death. Josie had the kind of mind that reminded him of himself fifty years ago. Sadly, he would never know the contents of the bowel or the stomach belonging to Jack Marrow.

'So when will you know who it is?' Josie asked.

'We'll do several tests and we might get lucky.'

'Were they murdered?' she asked.

Ted looked over his shoulder towards the terrace.

'I won't tell,' she whispered.

'Yes. We think he was.'

Chapter 8

Despite the late hour they'd gone to bed, Kelly was up at seven to bright sunshine, seeing to Lizzie and getting her ready for breakfast. She let Josie sleep. Often, she'd be down before anyone else, studying like mad at the table. But with her exams over, she now spent hours in bed, and Kelly thought she bloody well deserved it too. She was a hard-working kid and Kelly hoped she realised her dream of going to Durham. It would be odd without her around, though. They'd grown as close as mother and daughter could, and when Johnny had moved out last year during their separation, Josie had stayed with Kelly. Durham seemed ages away but it was less than two hours by car.

She put the thoughts to the back of her head and dressed Lizzie in the first outfit of her day. She was a messy eater, not because she was clumsy – or clumsier than any other eleven-month-old – but because she loved her food and because nobody made a fuss if she splattered it all over herself. The child giggled as her mother tickled her and took her downstairs. It was Kelly's favourite part of the day, when she had the house almost to herself and she could take her first cup of coffee out to the terrace and listen to the birds, and watch the gentle drift of the River Eamont flowing slowly towards Ullswater, which she could see to the west. The tap-tap of Lizzie's cup on her walker tray, as well as her singing and charging about over the wooden decking, was part of this routine now.

By the time Johnny came downstairs, showered and changed, Kelly was ready to hand Lizzie over and leave. She squealed and raced around the kitchen in her walker when her father appeared, and demanded the first kiss of the day.

They said their goodbyes and Kelly left the house and got into her car. Ted and Henry's cars were still parked in her driveway. They made a curious pair, like an old married couple. They'd shared stories, and bickered long after midnight about the pile of bones they'd found, each vying for the primary role in the investigation. Kelly wondered at the lack of knowledge she had about her father's life before she'd met him, well into his sixties.

She liked Henry and she could imagine the pair sharing a pokey flat in London, comparing notes and girlfriends. She'd seen something in her father light up when Henry was around, as if it reinvigorated his past life. She knew that if the skeleton was to cough up any answers this week, then it could be in no better hands than the pair of old friends. It would be a frustrating wait, though. In the meantime she had plenty of work to be doing, and she tapped her steering wheel as she listened to the radio.

All the chatter was about the skeleton.

Scores of theories were doing the rounds and she rolled her eyes as the station invited so-called experts on to the show to give their opinions. Some said bones couldn't survive under-water longer than ten years, others said there was an ancient settlement in the woodland at the foot of Helvellyn, and this was evidence of a long-awaited archaeological site.

She knew different. She was convinced by Ted and Henry's arguments, be they all circumstantial up to now, that Captain Jack Marrow was a homicide victim from the latter last century. The bracelet, amalgam fillings and leather restraints all attested to this, but the press had been told nothing.

As she approached Penrith, Kate called her from the office.

'The press are camped outside, you'll be mobbed when you arrive,' Kate told her.

'That's all right, I'm used to phones being shoved in my face. Any surprises?' Kelly asked.

'No, it's quiet in the office, unlike downstairs. We've had a couple of calls from relatives of missing persons on file, asking if it's their son or daughter. It's harrowing.'

'Tragic. We can't confirm or deny anything at this stage,' Kelly said. 'Loved ones of the missing always get a ray of hope when a new body is found. I wish I had some good news for them,' Kelly said.

Kelly had worked with enough victims' families to know that loved ones would rather have a dead body than a missing one. A person who simply disappears causes years of untold agony for those they leave behind. It wasn't uncommon for families to wait decades for answers, and some never got the closure they so desperately sought from being able to bury, and say goodbye to, their relative. It was one of the more sorrowful elements of police work.

'Is anyone else in?' Kelly asked.

'Just Emma. We're working on the files of missing persons in the last twenty-five years, in between our other stuff.'

'We might have to go back even further,' Kelly said.

'Really?'

'I spoke to Ted last night after he'd finished with the anthropologist. They both came over to mine. They're thinking latter twentieth century. But let's start with the cases up to the last twenty-five years and we'll work backwards after that. I don't want us spending too much time on this at the moment, because they were both hopeful that, with the fillings in the teeth, we should get a positive dental match on the database.'

'Sure. Well, we've come up with three young men so far who match the brief – late teens to early twenties – and we've got all the files prepared here for you to look at when you get in.'

'Right, I'm five minutes away.'

They ended the call and Kelly swung past the castle and approached Eden House. Kate had been right. Thanks to several high-profile cases over the last couple of years, Kelly's face was instantly recognised and the bastards even knew her car, though she'd changed it in the new year from an Audi to a Merc. She'd loved her old Audi, but the Merc's controls were similar and, after she'd got used to it, was a dream to drive. She'd gone for a GLA SUV and it negotiated the lanes of the national park admirably. But the press knew this vehicle too. She tutted as they blocked her way in to the car park. She ignored them and waited until three uniformed officers cleared the way. She could hear some of the questions though the glass.

'Kelly, was the victim murdered?'

'Kelly, will you dredge the reservoir?'

'Kelly, can you tell us who it is?'

'Kelly, when will you release a statement to the press?'

But it wasn't any of these inquiries that made the tiny hairs on her arms stand up.

The sun shone through her windscreen as she finally cleared the throng of journalists and parked around the back, safe from prying eyes. She didn't want anyone to know that she was rattled. She supposed that she'd have to get used to the incessant theories being thrown her way until they had a positive ID on the skeleton, and she hoped it wouldn't be long.

It was something quite different and unexpected that unnerved her.

A memory. A school trip. Decades ago. It was the last summer before their GCSEs, they were sixteen. The girl disappearing in the middle of the night, and turning up dripping wet and freezing, and the teachers telling them not to gossip. She tried to recall names and faces but she'd buried them long ago, and they'd all moved on. Why couldn't she remember their names? They were part of her past, and forever entwined in her history, her timeline. Why had her brain decided that they were forgettable?

Was it because of what happened next?

Chapter 9

Michelle Parkinson opened the office at nine a.m. sharp. She wore wellington boots, despite the heat, and her job affording her the luxury of rolling out of bed in a T-shirt and shorts every morning suited her well. No fuss, no make-up, no shower, just work to keep her head and hands busy. She'd started renting caravans from her auntie's farmland back in early 1999. Education wasn't her thing. She'd dropped out of A Levels mid-way through. She belonged in charge of something. No boss. No routine. School had made her feel like a caged animal, or a circus act, expected to perform for others, and she didn't understand why people stuck with it. The outdoors called to her. That, and making enough money to survive on her own. They were her only masters. Both gave her freedom.

The month of July was fully booked and she greeted new arrivals with the same enthusiasm as she waved farewell to others who promised to return next year. Bookings were healthy. She'd moved on since her time renting three battered old vehicles in a field to owning forty wooden cabins, of varying sizes, in the grounds she called Parkie's.

She'd never married.

The dream of finding Mr Right had come to nothing. She put men off, with her fancy for independence and her sharp edges. Besides, she'd seen too many friends shackled to the impossible fantasy come crashing down in despair.

This morning she was suitably distracted, just how she liked it. She kept busy. Her mind had a tendency to wander and so she kept it quiet by exhausting her body. She'd learnt to fix pipes, install washing machines, fit roofs and paint fences in her twenty-odd years in the business. She relied on no one. People always let you down in the end.

She finished off checking out two guests and prepared to welcome a family of five, who were staying at the park for a whole week. As part of the service, she liked to plan walks and suggest days out. There was no one better qualified to do so, she'd lived here all her life. The hills around Keswick offered some of the best hikes you could wish for, as well as Whinlatter Forest trails, water sports down at Portinscale jetty, steamer rides, walks around Hope Park, afternoon tea at the Lodore Hotel, and other more structured packages. She got a cut from the local guides and eateries when she directed visitors their way, and she reckoned that she could retire early should she want to. Which she didn't. The last thing she needed was time to think.

'Thank you so much, come again soon!'

Her holidaymakers always left happy and gave her rave reviews on Travelzoo and Tripadvisor. Her ratings were a solid five stars.

She followed the last guest out of the small office and waved them off, just as a large Audi Q7 pulled in and parked close to where she was standing. A couple got out of the car. The man had been driving, of course, and a hackle raised underneath Michelle's T-shirt. The woman was a walking advert for the repressed, wearing slacks and a polo top, with perfectly coiffed hair, as if she'd been flower arranging all morning. The man strode confidently towards the office, straight past Michelle, who said nothing. Three bored teenagers poured out of the back doors and she smiled at them. Behind her, the man tutted and began berating his wife, accusing her of having the wrong check-in time.

'We needn't have set off so early,' he said. The woman cast her eyes downwards and Michelle looked at the kids, who were oblivious to how the man imposed his toxic masculinity on the family. They were used to it.

'Family Graham?' Michelle asked the woman, who looked up and smiled.

Michelle stretched her hand out and shook the mother's hand, who glanced over Michelle's shoulder, looking for her husband to introduce. Before she could, Michelle directed her to the office and began telling her the details of their cabin and the rules of the park, fiddling with keys and paperwork.

It didn't take the man long to appear behind her, his eyes flicking suspiciously around the office, looking for something wrong with it. Michelle got the impression that it was the wife who'd booked the holiday park.

'This is my husband,' the wife said meekly.

'Mr Graham,' Michelle introduced herself courteously. 'We're just signing you in.'

'Do you want my signature?' he said arrogantly.

'No, your wife's is more than good enough.'

'My card?' he added. The wife blushed slightly.

'Thanks,' Michelle said.

The wife stood back and he checked his pockets.

'Ah, my wallet's in the car, go and get it will you?' he said to her.

Michelle held her breath, willing the wife to tell him to get it himself, but she obediently walked out to the car and found his wallet, bringing it back to him. He handed the card to Michelle.

The kids were moaning.

'Long journey?' she asked the wife.

'Too long.' Mr Graham answered for her. 'British roads. We thought we'd try a staycation this year, so we're missing out on the Bahamas. The kids aren't too happy.'

'Well once the Lakes gets in your blood you'll come back year on year. They'll love it, you'll see.' Michelle directed her encouragement to Mrs Graham but she remained silent, always waiting for her husband to speak first.

Michelle handed keys over and told them she'd walk to their cabin if they wanted to drive so they could unload their luggage. She reached under her desk and unclipped her two golden retrievers from their baskets. Thelma and Louise wanted to check out the new visitors and the kids giggled and petted them, distracted.

'Want to walk with me?' she asked Mrs Graham, who looked at her husband, who shrugged and walked out, getting in the driver's seat and telling the kids they could walk too.

It wasn't far, but long enough for Michelle to feel Mrs Graham's sadness. Michelle always wished she could invent an algorithm to wheedle out character flaws at the time of booking, but she hadn't come up with one yet. Dickheads like Mr Graham paid her bills.

They arrived at a fine cabin, beside the man-made lake that had been there when Michelle bought the park. It had been a sorry lagoon then, overgrown and neglected, and she'd transformed it. Mr Graham parked in the spot outside the entrance and he got out and sized up the place.

'It's good to rough it sometimes,' he said, going to the boot to retrieve cases.

'Get over here, you lot, and take a bag,' he barked at the kids.

Thelma and Louise avoided him. They were Michelle's best detectives when it came to humans. The dogs wandered off in search of calmer auras to hang around, and the youngest kid followed them; he looked around thirteen. It wasn't long before they started barking and the sound made her smile. That and the silence of the mountains were what grounded her, for sure.

'Do they do that all day?' Mr Graham asked.

'No. Not normally. They must have spotted a cat,' Michelle said. 'I'll leave you to it, you should find everything you need inside. If not, you know I'm not far away.'

As she walked away she heard Mr Graham admonishing his wife for putting a heavy suitcase on top of a lighter one. Michelle couldn't help hoping that Mr Graham took a tumble down a gully in the coming week ahead. In fact, she had just the walk for them that a city ego like his might struggle with.

Thelma and Louise were still carrying on making a din when she got back to the office and her irritation, compounded by Mr Graham's bullying of his wife, made her seek them out. She followed the rhythmic barks and they led her to the cabin she'd let Jason crash in until he got on his feet again, three weeks max, he'd said. He'd been there four years. It was a rundown structure on the edge of the park that she didn't rent out, and besides, he was an old friend.

She could hear the teenager talking to the dogs and she went to the side of the cabin, listening to him chattering away. It was innocent and sweet. She waited beside the cabin and peered around the corner, where she saw him finding sticks for the dogs. But the two retrievers were more interested in something they'd found under the cabin.

'Here, here,' the boy said, patting his knees.

Poor thing, Michelle thought, he was desperate for a friend. He probably got lost in that family. He'd love it here. But still the dogs wouldn't come, and then the barking turned to howling. She rounded the cabin and startled the boy.

'Sorry!' he said, far too quickly. Michelle could see that he was used to being in the wrong.

'Hey, no worries, you've done nothing wrong, I just want to know why Thelma and Louise are going mad. It's not you, let me look.'

She bent down and peered underneath the cabin. She got out her mobile phone and put the torch on and swung it around, spotting Thelma and Louise standing over something. She tutted and called them. Then she got a whiff of something like fresh fish.

'You better go back to your mum and dad,' she told the boy. 'I reckon a deer or something has got trapped under there.'

'Can I see?' he asked, kneeling down and crawling underneath. She tried to stop him but before she could get hold of his disappearing ankles, the boy had cried out and banged his head. He rubbed it and shot out from under the cabin backwards. He looked like he'd seen a ghost.

'What's the matter?' Michelle asked.

'It's a man,' he said. The lad was shaking, but Michelle was calm.

'God, I bet Jason is drunk again,' she said, loud enough for the boy to look at her oddly.

She knelt down and shouted Jason's name, but there was no answer. Michelle was aware of other feet behind her as she crawled underneath the cabin on her belly. She recognised the voice of Mr Graham, who was admonishing his son for wandering off. He knelt down and shouted at Michelle.

'What's going on? My son said there's a body under there.' Michelle ignored him.

Jason wasn't drunk, and he wasn't asleep.

'Call the police,' she said.

'I knew it,' she heard Mr Graham say. 'Jesus, where the hell are we? What a shitshow.'

Chapter 10

Kelly was informed about the discovery of a body underneath a holiday cabin in Keswick at just after eleven o'clock. Ordinarily, a single deceased male wouldn't attract her attention, but one that had been described as 'fairly mashed up' by the two officers first at the scene strongly indicated that it wasn't a natural death, and merited her inspection.

She knew no other details apart from that the body had been found by the park owner's two dogs. She and Kate grabbed their bags and headed out of Eden House to face the photographers and reporters who were being held back by a temporary cordon that was wreaking havoc with the traffic. They offered no comment. One of the more bizarre stories circulating about Captain Jack Marrow was that he was in fact Elvis Presley.

They drove along the A66 to Keswick and turned off before the town, to Parkie's Cabin Park, where they found two uniformed officers trying to calm several people inside the main office. Kelly parked and she and Kate went in to the office. The uniformed officers were relieved to see them and told Kelly that two further officers had arrived and were standing guard around the crime scene.

They also told her that it was a fourteen-year-old boy who'd discovered the body, and he was being consoled by his parents. She was introduced to the park owner and as she shook Michelle's hand, the familiarity in her features dragged up long ago memories from when their faces were brighter and tighter, and looking forward not back. Underneath the rough clothing and the deep wrinkles on her face, and the greasy tangled hair, Kelly recognised her.

'Michelle, it's been years,' Kelly said.

They'd gone to school, and then sixth form, together. They'd shared English class. She remembered Michelle being aloof and aggressive, and Kelly had given her a wide berth.

'Kelly, you're looking fine, lass. I heard you moved away. I saw you on TV, proper up there with the law.'

The handshake was functional, but Kelly noticed the woman's strong and confident grip, which she admired. They locked eyes. She was bubbly and rather upbeat for somebody who'd had a dead body discovered on her property.

'We went to school together,' Kelly explained to Kate. 'So, you own this place?'

Michelle nodded. 'Yep. All mine. School was never my thing. I work with my hands.'

Bumping into old school peers who were suspicious of the one who got away was common, but it had ceased to bother Kelly years ago. However, memories of this particular group of contemporaries was unwelcome, because it included people she'd rather forget.

'Can we go now?'

Kelly swung around to where the question had come from and saw a man stood with his hands on his hips.

'Kelly, this is Mr Graham, the father of the poor boy who made the discovery,' Michelle explained.

'Ah, Mr Graham, where is your son?'

'He's being consoled by his mother. You can be damn sure we'll never come back to this place,' he said.

Kelly thought she saw Michelle's mouth turn up in a smug smile. She remembered that Michelle was never one for holding back, and a cocky southerner like this man, who cast aspersions on Michelle's business because of something utterly out of her control, was likely to rile her. It riled Kelly too.

'We'll need to speak to him. Don't worry, we'll be mindful of what he's been through, and in your own time, of course. Did he tell you what he saw?'

The man took a deep breath. He was the type of guy who waited until all attention was on him before he spoke.

'We'd been here five minutes and the dogs were barking, he followed them and found a corpse. What else is there to say? Welcome to the Lake District?'

Kelly ignored his ego's cry for attention.

'I get the impression you'll be leaving, then?' Kelly asked.

He huffed. 'How can we stay here now? Jesus, we're safer in London.'

'I'd appreciate you staying for the moment, we'll have to take a statement.' Kelly turned to Michelle. 'Can you show me?'

Michelle smiled and nodded.

Kelly followed her out of the office and towards a forested area. The park was pretty and Kelly was impressed with it.

'Kate, can you handle the family?'

Kate nodded and retrieved a notepad from her bag.

'Have you got any gloves?' Kelly asked.

Kate supplied her with a pair.

Two golden retrievers trotted beside her and Kelly patted them. 'The detectives?' Kelly asked.

Michelle nodded. 'Thelma and Louise.'

'Great names. I love that film. Two strong women.'

'The only type I want around me.'

They walked on in silence. Michelle pointed to a cabin on the edge of the park, slightly away from all the others, and one that didn't look so grand. A beautiful lake sat serenely behind it. Her first thought was, why kill somebody and not use that to hide the body?

'He's under there. I let him live here rent free. He had his troubles, poor sod. My guess is he picked a fight with the wrong person, drunk likely,' Michelle said.

'You knew him?'

'Of course, don't you know? It's Jason Cooper.'

Noises from a past deeply hidden in her memory fought for attention in Kelly's brain. The name. The face. Clarity assaulted her and she recalled that Jason Cooper had also been in her school year.

'Didn't you two have a thing?' Kelly asked.

Michelle nodded. 'Years ago. He got married and had a couple of kids. It ended, I offered him a place to stay. We weren't in a relationship. I was doing an old pal a favour.'

'I see. That's very kind, Michelle. You looking out for your mates.'

'It's rare, some people only think of themselves.'

Kelly tried to remember Jason's face but it had been years, over twenty for sure, since she'd seen him. Suddenly the heat of the approaching midday sun hit her and she felt hot under her T-shirt. She popped her sunglasses on, as if that would protect her from all the memories she was about to revisit. Her earlier paranoia about another of their classmates assaulted her and she felt a rising tide of puzzlement that seemed to be getting more overwhelming by the minute.

Snapshots of their classroom. The school trip. Michelle as a young teenager, always hanging around with the boys. Kelly had envied her. Her confidence, her swagger; that secret something that can't be bought, and that most kids wished they had.

Michelle and Jason had been quite a thing back in the day. They waltzed around sixth form like they owned the place, until they both left suddenly. Even then, Kelly had wanted somebody to bewitch her like that, but not anymore. It was one of the reasons she'd moved away; she'd never fitted in.

The coppers keeping guard at the cabin nodded to her and then glanced at the dogs.

'Can we keep them away? Did they remove anything, do you know?' she asked Michelle.

'Thelma and Louise? I didn't see them with anything, I would think they just had a good sniff.'

Michelle called the dogs and they went to her. Their obedience was impressive. She produced a double lead and tied them to it, walking them to a fence close by and tethering them to it. They sat down in the sunshine, panting.

Kelly bent over and peered under the cabin, slipping on the plastic gloves with a pop. The smell of death was definitely in the air, and this weather wouldn't help, but it wasn't overpowering, which told Kelly that he'd not been dead long.

'How did you know it was Jason?' she asked.

'I recognise his boots. He works at Connor's farm, and he always leaves them outside his cabin. And also his hair, and the tattoos on his arm. It's him.'

'And was he in the habit of falling over drunk? An accident perhaps?'

'He was an alcoholic, Kelly. We didn't all do as well as you.'

Kelly noticed the first flicker of emotion and Michelle's self-assurance dipping.

'Right.' Kelly looked around. 'Thanks, Michelle, I'll take it from here. It'll take me a while to have a good look around.'

'That's fine, I'll wait. I want to know what happened as much as you do.'

'Of course you do.'

Kelly took in the scene. The uniformed officers had had the wherewithal to secure a taped boundary ten feet clear of the whole perimeter around the cabin. She walked the whole line, bending down, looking for detritus that might indicate a struggle. At the rear, by the door, she saw evidence of recently disturbed earth and saw that a drag mark led under the cabin. She knelt and focused on three or four large stains of brown deposit that looked to her like dried blood. She took markers from a bag she'd brought from the car and placed them at each spot. Then she measured them. There was one smear that was larger than the others and when she inspected it, it appeared that just the surface of it was dry. It had the consistency of a blob of the home-made jam that her grandma put in the fridge to test

for the setting point, with a film of skin on top. She followed
the smear and saw that it disappeared under the cabin. Her eyes
followed the drag mark backwards and she noticed that it led to
the step up to Jason's cabin. There was what looked like dried
blood there too and she marked it.

She lowered her body further and crawled a little way under
the cabin, avoiding the stains. As she got closer to the corpse, the
smell of putrefaction grew stronger. It didn't take long for the
first signs to show, she knew from grim experience. She logged
the details of the position of the body, and the orientation of
the cabin, in her mind. She looked around the perimeter of the
cabin, to where she saw the feet of the coppers and Michelle,
and the dogs staring at her from a distance, eager to get involved
and do their job: retrieve. She looked at Jason. He was a big
bloke. He'd been tall, she remembered, even at seventeen. His
appearance was coming back to her. She saw the tattoos on his
lifeless arms and the mass of golden hair mingled with dried
blood. He was face down, and she could see that his shoulders
were broad and his legs muscular. In the cramped space, it was
difficult to work out how anyone could have manoeuvred such
a lump under here.

She crawled a little further in and saw the state of his head
and, as her eyes accustomed to the dim light underneath the
space, she saw that the blood pattern followed the drag mark
and stopped at his head. This is what the officers had seen
when they reported to her that he was 'mashed up'. A small
glimmer of silver caught her eye and she fingered the earth
around the spot until she felt something hard and crawled a
little further into the cramped space. Next to the body, close to
his face, she saw a silver chain, and attached to it was a locket.
It wasn't a man's, but it could have been under here for years.
It was old fashioned, like something her mother's generation
would wear.

Until they could corroborate Michelle's ID of the body,
she would hold tight on informing next of kin. But there was

no doubting at all that the man was dead, and even if he had been inebriated, there was absolutely no dispute that he'd been murdered.

Chapter 11

The boy was traumatised. But it was more because of his father's interference and stress than anything else. He simply wasn't helping. The mother was softer.

When Kelly got back to the park office, Kate rolled her eyes and asked for a private word.

'He's leading the boy and making him absolutely terrified,' Kate said.

Kelly nodded. 'Is he in there?'

'Yep, the mother brought him back to the office. She seems extremely stressed as well, and I don't think it's just the circumstances. His name's Ben.'

Kelly went into the office and introduced herself to Mrs Graham, who was pale and nervous.

'Hi Ben,' she said to the boy. He looked at her and to his father, then nodded.

'Mr Graham, can we have a minute? One parent is all we need to comfort your son at this time. I'll be quick,' she said.

He glared at her and stormed out of the office, slamming the flimsy wooden door. She heard Michelle outside.

'It's hardly the door's fault, pal. Show some respect,' Michelle said. Kelly, who was still stood up, peered through the window, and saw Kate standing with Michelle. Mr Graham squared up to both of them.

'If it wasn't for you, we'd be miles away from this shithole, on a plane, without my son seeing some dead man under a shitty caravan,' he snarled.

Kelly's gut tightened. She hated bullies. But she also didn't fancy anyone standing up to Kate Umshaw or Michelle Parkinson. He'd have no chance.

'There's no call for that, sir,' Kate said.

Michelle stood behind her, grinning.

'Better go find that plane,' Michelle goaded.

Mr Graham took a step towards her, then, blocked by Kate, thought better of it and stormed off.

Kelly turned back to Ben and sat down on a chair. Mrs Graham held her son's hand.

'We can provide all the help you need to come to terms with this awful thing you've seen, Ben. If you feel able, could you tell me exactly what happened this morning? It could help.' She turned to Mrs Graham. 'If you decide not to stay, we can arrange for you to check in somewhere else. It's understandable. Where do you live?'

'Kensington.'

'I can arrange for continued support when you get home,' Kelly said. 'I used to work for the Met, I know their victim support unit is excellent.'

The London link threw an olive branch to the mother and she breathed a little easier. They chatted about the city, and Ben told her about his school. He also told her how he knew that the man was dead, because he'd smelled the same sweet whiff of expiry in biology when they dissected a baby pig.

He told her that the owner of the park, Michelle, had followed the dogs and saw what he saw, and had comforted him.

'She's nice,' Mrs Graham said.

Kelly was as gentle as she could be with the boy, and he dictated a statement to her with admirable maturity. After they left, Michelle returned to the cabin and made them tea.

'Do you think they'll stay?' Kelly asked.

'No idea. I won't charge them if they do. It's not an ideal way to start a holiday, is it? It looks bad for the park.'

'We'll try to keep the press at bay for you. So, tell me about the last couple of days. When was the last time you saw Jason? How was his mood? Did he have any—'

'Enemies?' Michelle jumped in. 'Plenty.' She took a deep breath. 'Jason was a lost soul. He thought he could get by in life by working hard and providing for his family. Unfortunately, he didn't realise that he also had to do all the other stuff, like be a husband and a dad. He was useless at both. Oh, don't look at me like that, Kelly. He was. He said it himself. Anyway, we remained good friends all these years and he often used to come and crash in a spare cabin when he'd been kicked out after an argument. He told me stuff, you know, like old friends do. He got into trouble. He couldn't hold down a job. He began to drink more and more. I guess when you get into your forties, you start to consider if your life is going anywhere. I see you've done well. But then you always kept your head down and your hands clean, didn't you?'

Kelly went to reply. She felt criticised and her defence mechanisms rose up inside her.

'Oh, I'm not judging. It's a compliment. Some kids know the importance of studying hard at school from a young age. I wager that those three kids out there, with that arsehole of a father, all study hard and will go on to accomplish great things. Ben will get over what he's seen. Jason and I were rootless. That's all I'm saying. We learned the hard way, you know. Bad start. Bad luck. This was inevitable, I guess.'

Kelly didn't know what to say. Michelle made it sound all so inescapable, that a child's destiny is written from the word go, and there was little you could do to change it. It was depressing, but Kelly knew from experience that she'd hit the nail on the head. She saw it all the time: career criminals coming from the same backgrounds of neglect as Jason Cooper. His parents were dead, Michelle told her; his father had been a drunk who hit his wife and kids, and his mother a slight meek woman who kept the family secrets.

'He was just a big kid really. I never judged. I just provided him with a roof over his head.'

'And the last couple of days?' Kelly asked.

'He had some falling-out at work, over a fence he laid wrong. He got into a proper fist fight with a farm hand, who accused him of pilfering. Connor's farm will be able to tell you more. He was probably tanked up. He often stopped at The Swan Inn for a couple – or more than a couple – on his way home from a shift, to unwind, you know? Maybe he was followed? Jason would never have lost a fair fight, and it's the back of his head that's damaged isn't it? I didn't hear a thing, though. I was busy with the environment people, trying to tackle the Japanese pondweed that's invaded my lake. I was with them until about five-ish yesterday. I saw Jason yesterday, about eight in the morning, on his way to work. He waved and we'd said we'd have a drink maybe at the weekend. Sometimes I did his shopping. I dropped some off on Friday. I'm babbling now.'

Pain passed across Michelle's face and Kelly gave her a moment. She'd lost a pal of over twenty years and despite her bravado, he meant something to her, she could tell.

'That's really helpful. It means he was killed yesterday. I don't suppose you have CCTV?'

'No, sorry. I keep meaning to install it. Maybe I will now. The front gate is always open too, because I have visitors arriving all hours.'

'I'll need a complete guest list of everybody on the premises yesterday.'

'Of course, I can print a list off for you.'

'We'll have to search the whole park. We might get lucky and find a weapon. We'll also have to search Jason's cabin.'

'Of course. The door was always open.'

'Really?' Kelly asked.

'Nobody expects anything like this,' Michelle said. 'I'm going to start locking mine and I'm sure my guests do anyway.'

'Good idea. Where do you stay? Here on the park?'

Michelle nodded. 'Over there,' she said.

Kelly saw that she pointed to a large cabin not far from the main entrance.

'Did you see Jason come back from work yesterday?'

'No, like I said, I was down at the lake. Then I went into Keswick for the market. I was knackered. I had an arrival at eleven last night, then I put the dogs to bed and closed up.'

'Do the dogs roam free during the day?'

'Pretty much. They don't wander off.'

'And you didn't hear them barking particularly last night?'

'No, and they're my guards. If somebody came inside the park and killed Jason last night, then Thelma and Louise didn't hear it.'

Chapter 12

A black coroner's van made its way through the park and was directed to Jason Cooper's cabin. It was a sorry sight, and not one which was usually associated with holidays. The news had got out, but thankfully interest was still concentrated around the skeleton remains from Thirlmere. Journalists and reporters had now migrated to the Penrith and Lakes hospital, where the skeleton was being examined. Details of the murder of Jason Cooper were kept under wraps and local gossip was slow to filter through about a drunken man who'd been discovered under a cabin. The news Kelly heard was that a homeless man had crawled under there for warmth. A few local reporters had tried to interview Michelle, but she'd stuck to the script and refused. They'd moved away.

For now, the park was relatively quiet.

Their inquiries centred around whether anyone had heard any unusual commotion at the park last night and accounting for everybody's whereabouts. So far, all those interviewed had solid stories and gave no impression that they knew the man living in the cabin at the edge of the forest by the lake.

Thanks to Michelle, they had a timeline. Jason had been alive yesterday morning, and so that made his corpse fresh and the time of death pretty easy to ascertain. He'd soon be on the coroner's slab, and Kelly felt a pang of guilt for making her dad so busy. However, she knew that Henry would be taking the lead on Jack Marrow, so Ted would be thankful for the work, because she knew he'd be chomping at the bit to get involved.

She'd called him and warned him that a cadaver was on his way and this time, it had flesh on the bones and they had reason to believe that he'd been alive up until sometime yesterday afternoon when he finished work.

Kelly had checked with Connor's farm, and Jason had clocked off at four in the afternoon. She'd also called The Swan Inn, and he'd been drinking there until six p.m.

She and Kate agreed that their first priority, apart from confirming Jason's ID, was interviewing the current staff list at Connor's farm, and finding out with whom Jason had his falling-out, and why.

Kelly went to find Michelle and found her sat on a bench by the lake, near the wooded area, close to Jason's cabin.

'Hey, are you okay?' she asked.

Michelle nodded. 'I put this here when I bought the place. It's where I think.'

'It's nice. This is some property you have. You must be very proud.' Kelly sat down.

Michelle smiled. 'I've done okay. Do you ever think about those times, Kelly?'

'What, school?'

Michelle nodded. 'Our class, where they all ended up. Dave Crawley in prison, and I heard you were to thank for that, piece of shit he was. Trafficking women is about as low as you can go. Now Jason. And Brian.'

Kelly realised that Michelle had given her the missing piece to the puzzle that had been bothering her all morning. Brian Miller. Their classmate who'd disappeared all those years ago. They never found him. It was a sorry group of alumni.

'You ever see the others?' Kelly asked. 'Paul? Carol or Tracey?'

'Nah. I don't believe in looking back. Digging up the past is something that should be avoided. Maybe that's why Jason got into trouble, he was always pining for what was. He had a fling with Carol, I know that.'

'Do you think it's worth tracking them down? You never know with cases like this, old scores and all that.'

Michelle stared out at the lake. 'You're the only one who left. You did the right thing. I know Jason drank with Paul sometimes, especially after Dave was sent down. It hit him hard.'

Paul 'Flash' Gordon, Jason, and Dave Crawley were best pals at school, and sixth form. Maybe it was worth looking into. She gave Michelle her personal mobile number.

'If you need anything,' she said.

'Maybe we should hook up. You know, for a drink?' Michelle said.

Kelly was surprised but at the same time flattered.

'I'd like that, Michelle. I'd like that very much.'

She left Michelle to her musing, and went to the cabin where the coroner's van was loading the body. She'd instructed the forensic team to get started inside and outside the cabin, and, so far, they'd found nothing of any real consequence: the odd fag butt, some litter and the like, but no weapon. They only had the locket that looked out of place for the crime scene, and that had been bagged and tagged as evidence. They were taking blood samples from the drag scuffs, and searching for any clues that a second person might have been inside the cabin with him. Michelle had gladly given over her fingerprints because, as the landlady, they'd be all over this place.

As the body was removed from under the cabin, ready to be placed into a black body bag and transferred to the mortuary at the request of the coroner, Kelly stopped them and peered at the face of the man. It was dirty and pale, and apart from the obvious, he looked asleep. She recognised the shape of the face and the curve of his mouth. It had transformed over the twenty years since she'd last seen him, but it was also hauntingly familiar to her because he'd been her crush for two years, and she'd stared at him at every opportunity. But he'd preferred the dangerous girls. The ones who threw caution to the wind and got into trouble with him. They smoked, scraped shapes into their forearms, walked for miles around lakes late into the night, and experimented with drugs and alcohol. Kelly Porter was too pristine for this likely lad.

She stared at him and a lump formed in her throat.

It was the face of Jason Cooper, and she had all the confirmation she needed.

Chapter 13

The Swan Inn wasn't busy. It was the type of place that welcomed farmers and workers with muddy boots and tobacco pouches. It was grimy underfoot and Kelly was hit by a stench of cigarettes and stale alcohol as she entered.

'Nice gaff,' Kate said.

Kelly grinned.

Of course, by law, smoking was no longer permitted inside public premises, but in the middle of the countryside, with sheep farmers and forest labourers wanting only to swap stories and eat crisps, getting away from their exertions until they had to go home, who was going to report them?

A young woman sat behind the bar, looking bored, and didn't look up when they entered. It was mid-afternoon and a few men sat alone, sipping their pints. The TV threw out pictures but no sound, and the doors were kept open, presumably to get some air into the place. It didn't work. The atmosphere inside was as stuffy as it was in the rising heat outdoors.

'Afternoon,' Kelly said to the young woman behind the bar, who, on closer inspection, appeared to be a teenager.

'What can I get you?' the girl said, slipping off her stool. 'You're not regulars.' It was a statement, not a question.

'We're all right, thanks, we just want to ask you a few questions about a customer of yours,' she said, showing her lanyard.

The girl stiffened when she glanced at the ID.

'Don't worry, you're not in trouble. What's your name?'

'Connie. I just work here, I can call the manager if you like?'

'Were you working yesterday, Connie?' Kelly asked.

She nodded.

'You know Jason Cooper?'

Another nod.

'In here yesterday was he?'

'I don't want any trouble.'

'You're not in trouble.'

'Is he?'

'We're not at liberty to discuss that at the moment, we just want to know what time he was here and what he ordered.'

'Right. Well, he came in here after work, as he always does, about four in the afternoon, and I think he left about six-ish. He drinks Lakeland Pride, sometimes with a couple of whiskeys.'

'Sometimes?'

The girl walked away when she heard a beep and Kelly assumed it was the glass washer. She was correct, as the girl bent over and opened a door, letting a plume of steam escape into the air. It didn't help the ambience, but provided the girl with a hiding place.

'How many pints of Lakeland Pride did he have yesterday, would you say?'

'Two.'

'Two?'

'Maybe three.'

'You'll have till receipts.'

'Not with names on.'

'Was he drunk when he left?'

'No. I've never seen Jason drunk.'

'Really? He's a moderate drinker, then?'

'Not moderate, but used to it. He can hold his booze.'

Michelle had told Kelly that Jason was a hard drinker; this would fit with him being able to pull off several pints in the late afternoon, and a young woman like Connie thinking he could handle it.

'So, three pints, then.'

Connie stood up and smiled, chewing gum. 'Like I said, he wasn't drunk.'

'What mood was he in when he left?'

'He was giving me a lecture about making sure I get my education. He was funny like that. I guess he didn't want me to make the same mistakes he did, but I won't.'

'You're studying?'

'Yep, I finished my A Levels in June.'

'Like my daughter.'

'That's nice.'

It was disingenuous. But Kelly wasn't after connection, just a way in. 'So, his mood?'

'Usual. What's this about? Has he done something?'

Kelly ignored the question. 'Any rifts or squabbles in here involving Jason?'

'We don't get trouble in here,' Connie said.

Kelly found that hard to believe, all pubs saw their fair share of combative behaviour.

'So, he came in at four, drank three pints and left at six. Any talk of falling out with anyone?'

'He'd been attacked at work. I saw his face. Some guy had punched him. It was a labourer, he called him Danilo. He's trouble apparently. He said he was being let go because of it, it wasn't Jason's fault, he didn't do anything. I know him. He wouldn't hurt anyone.'

Connie was earnest, and Kelly almost believed her.

'Do you know how he got home?'

Kelly knew full well that Jason had driven his car because it was parked beside his cabin, and that's how he got to work. Michelle said it wasn't there in the morning.

'Dunno. It was busy yesterday.'

'On a Tuesday afternoon?'

'Could he have walked?'

'Back to Keswick?' Kelly asked. From The Swan Inn, Keswick was a two-hour walk, but a ten-minute drive. The pub was nestled off the busy A66, close to Scales at the foot of Blencathra, and Kelly had parked there plenty of times to hike. Connor's farm was just a little further down the lane. It was fertile farming land around the plains of Skiddaw and Blencathra, fed by mineral-rich water from Bassenthwaite and Derwent. Kelly knew the Connor family by reputation.

'Look, a customer's decision to get into a car after three or four pints is their liability, not yours. I'm not interested in that, I just want to know how he got home.'

'I think I saw his car.'

'Thank you. And he left alone?'

Connie nodded.

'And, apart from the bruise on his face from a punch, at the hands of Danilo the labourer, he was unharmed?'

Another nod.

'Did he talk to any other customers in here yesterday?'

'No, everybody was watching the TV. You know, about the body in Thirlmere?'

'Yes, we know. Thank you, Connie, we'll be in touch. Anything else you remember about yesterday or anything that Jason said, then give us a call.'

She handed Connie a card.

They left. Kelly noticed Connie staring at Kate, and that was the whole point. One copper staying quiet always unsettled folk, and they were much more likely to reflect on the subject at hand if they felt uncertain about what the police thought about it.

—

Connor's farm was just up the road, across the little stone bridge over St John's Beck. It was one of the larger farms in the area, and employed countless transient workers from all over the world, as farms had to in order to survive these days. It wasn't lost on Kelly that Danilo wasn't a local name, but she was also wary of the potential xenophobic undertones of blaming a foreigner for misdemeanours.

They found an office and banged on the door. It was empty. A man strode across one of the yards, dressed in casual dirty clothes, with large boots making a stomping sound underfoot.

'Excuse me!' Kate shouted. They caught his attention and he told them that the owner was up at the house. They got back in the car and drove to the main house and parked outside.

Mr Connor was a large man, with friendly eyes and a warm handshake.

'My son takes care of the staff. I'm slowing down. I know Jason, he's a hard worker, but he needs to go easy on the ale, if you know what I mean.'

Kelly and Kate both smiled knowingly.

'What about Danilo?'

Mr Connor thought for a while. 'Is he a tall chap?'

'Not sure, we've just got a name,' Kelly said.

'Kev!' Mr Connor shouted indoors and invited them in.

The farmhouse was cool and the kitchen smelled of fresh bread. It was like stepping back in time. A younger man, probably in his thirties, ran downstairs and came to his father's call, and stopped when he saw the two female guests. His father introduced them.

'They're asking about Jason,' Mr Connor said. 'And do we have a Danilo on our books?'

Kelly realised that for all his sorrow over slowing down, Mr Connor was sharp.

'Is this about their fight? I've dealt with it. Has somebody made a complaint? It's not like Danilo to make a fuss,' Kev said.

'No, we're just enquiring about Jason's whereabouts yesterday. We were told that he worked a shift here, and got into a fight.'

'Is Jason all right?' Mr Connor asked.

'I'm afraid we're not able to discuss Jason's involvement at the moment, but we wanted to establish what he was doing here yesterday.'

'It sounds serious,' Mr Connor said.

'What was the fight about?' Kelly asked.

'Danilo said he'd stolen some tools, but it looks like he was mistaken, they turned up yesterday evening. But, well they got into a fight and Jason was thumped pretty good until somebody got in between them.'

'What about in the past? Have you had gear go missing? Do you do stock takes?'

'I'm afraid we're a bit lax in that department. My dad handed that all over to me a few months ago and I did see some gaps. I can dig out the notes if you like?'

Kelly nodded. 'Is Danilo at work today?'

'Yep, he's in the west field. I can take you to him in the tractor, but I can only take one, and I'm not sure you're dressed for it.' Kev glanced at Kelly and Kate's attire.

'That's fine, I'm sure I'll cope. Can we go now?'

Kev shrugged and led the way out of the farmhouse.

'Kate, you okay to wait here?' Kelly asked.

'I'll make her a cup of tea,' Mr Connor said generously.

Kelly followed Kev and he led her to a huge tractor in the yard.

'Can you climb up here?' he asked.

She was glad she'd worn trainers again today, knowing she'd be out and about. She hopped up to the cabin deftly and Kev looked impressed.

'Local?' he asked.

'More used to rocks, but tractors will do,' Kelly said.

They drove slowly out of the yard and Kelly couldn't help imagining what farm life must be like at this pace. Everything slowed down, nothing could be done hurriedly. They jerked and swayed and entered a field. Kev pointed to the area where he said Danilo would be.

'So what's this all about?' he asked.

'I'm not at liberty to say yet, I'm afraid. We just want help with our inquiries at this point.'

'I'll hear on the news then, is that what you're telling me? Jason okay? He didn't show for work today.'

'Is this all your land?' Kelly asked.

Kev got the message and changed the subject, pointing out the Connor boundaries to her. It seemed that the farm stretched as far as the eye could see. It was enormous.

'What's Danilo like?' she asked.

'Quiet chap. That's why I didn't lay him off for thumping Jason. It was out of character and they shook hands after.'

By the time they reached a group of labourers toiling in the heat, Kelly was quite getting used to the ride and her heart had slowed. She reckoned that farm life was good for the soul.

Kev stopped the tractor and shouted at one of the men, who'd all stopped work when they saw their boss.

'Danilo!' Kev shouted.

A tall man in muddy shorts and a T-shirt, standard summer farm attire, strode over to them, taking off his heavy gloves, and Kelly asked him some questions to get the measure of him. She had to look up to him as her trainers sank in the uneven field, but he struck her as a benign character, if somewhat closed and broody. He told her that he was at work until gone eight p.m. last night and he went straight to his caravan on the farm. His alibi was his two bunk buddies, and he hadn't seen Jason since their spat.

His version of events would need checking out.

'So, you didn't see Jason after he clocked off at four?'

Danilo shook his head. He stretched his back and wiped his brow.

'Do you have a grudge against Jason Cooper?' Kelly asked.

Danilo shook his head. His English was perfect, despite it being his second language. He'd followed work from his home in Portugal and he was happy to chat, though Kelly suspected that it was so he could get an elongated break.

'Is he all right? Am I under arrest?'

'No, there have been no charges filed. It's an unrelated incident, we just want to know if you saw him after work yesterday.'

'No. Like I said, I worked late and then ate and went to bed, after we watched Netflix.'

'Thank you, I'll get the details of the men you've mentioned from Kev. I'll let you get back to work.'

'It wasn't my fault,' Danilo said.

Kelly turned back to him.

'What wasn't your fault?'

'He was drunk, he called me a word I hate. He said I was lying about the tools.'

'Which tools did you think went missing?'

'I thought, and I still do,' he glanced at Kev, 'that he takes stuff home.'

'Like what?'

'Spades, hoes, equipment. I think he sells them.'

Kev sighed and put his hands on his hips.

'Thanks Danilo, back to work, mate,' Kev said.

They got back into the tractor cabin and this time Kev didn't offer to help her.

'What do you think to that?' Kelly asked him.

'He's probably right. Jason's time with us was coming to an end, to be honest. I know he steals stuff from us, and this was the last straw. I was going to let him go today.'

Kelly recalled the wounds to the back of Jason's head. A spade would do it.

Chapter 14

The job of scraping earth away from bone was a task that not even Ted's patience could withstand. However, Henry's work had paid off. So far, they'd revealed all of the iliac crest, the pubic bones, one complete femur, fibula, most of a spine, the skull, and two almost complete arms, with just both ulnas missing. The earth around the skeleton had been carefully sifted, bagged, tagged and sent off to a pedologist. They were hoping that the soil expert might be able to tell them how the body had settled on the bottom of a lake so well intact. They could also confirm or disprove Henry's theory that the body had been buried, though how that might have come about was a puzzle.

'It's likely that the body was weighed down and settled quickly in the mud at the bottom of the lake,' Henry said. 'Otherwise, we'd have a completely disparate set of bones. The fact that they've remained together so impressively suggests they rested on the lake bed pretty quickly after being dumped. We'll know more when the soil is examined,' he said.

It was abundantly clear that Jack Marrow wasn't killed underwater, however, without a pair of lungs to examine, or any organs at all, it meant that the cause of death was almost impossible to ascertain.

'Here we go,' said Henry, holding up Jack's skull. 'I'd say with the rate of the epiphyseal fusion of the skull and facial bones in teenagers, as well as the protruding brow ridge, this is consistent with a male, aged sixteen to twenty-one. Look at the mastoid processes.'

Henry had already begun the process of removing the adipocere, by soaking the long bones that he'd retrieved so far in a solution of warm water and typical non-bleach kitchen detergent. The process would take several hours and only then would they be able to piece together the skeleton properly.

Ted knew that an incredible amount of information could be gleaned from a skull. He also knew that the human skeleton wasn't properly fused until about the age of twenty-three. He was no anthropologist; he dealt in tissue, not bone. They were very different disciplines. But he did understand enough. The length and breadth of the skull, as well as the eye orbits and nasal aperture could all be inputted into software that would feedback pretty accurate conclusions on age, gender and ethnicity. The mastoid process was a chunk of bone above the jawline which played a structural role in the face. Those of a female were much smaller and more delicate.

Separating the head was an image that Ted was unfamiliar with. His autopsies usually left it where it was. Sure, he regularly sawed into the brain and spinal column, but to have the head completely separated, jarred him. He imagined Hamlet's gravedigger scene from the Bard's famous tragedy. Shakespeare had included the scene for some much-needed comic relief and it was appropriately timed, because it was having the same effect now. Yorick's skull had lain in the earth for three and twenty years, which was not far off their projected assumption for the remains of Jack Marrow. Ted half expected Henry to begin waxing lyrical on the question of humanity, but the vision soon dissipated. The moment was sobering, which was exactly what the poet had intended.

'Come and have a look at this, Ted,' Henry said.

He walked around the gurney. The bones were laid out, as if they were part of a child's jigsaw puzzle, and were beginning to resemble a human form, barring the ones being soaked. He peered closely at the skull, which Henry still held up like a child who'd found the biggest egg on the Easter hunt.

'Look at the meningeal sulci, it's quite stunning.' Henry said.

Ted nodded at Henry's reference to the tiny grooves made by the arteries that criss-crossed the inside of the skull. The cranium and parietal bone was intact but of course the mandible was missing, sent to the lab for dental ID. DNA could also be extracted from the centre of the largest molars, should they need it. But it was the rear of the skull that Henry was concerned with. He pointed out where the occipital bone was damaged. Henry rubbed away some dirt and dust, one tiny thumb sweep at a time. He indicated the mastoid process on the left side of the skull, but then traced a messy fracture all the way around the temporal bone, which left a gaping hole the size of a fifty pence coin. The tiny fanglike styloid process on the left side was also missing. It was a delicate bone but well protected in life.

'That's non-survivable and not something that a disease would cause,' Ted said.

'Exactly, and there are no fish in Thirlmere capable of such an attack, I assume?'

Ted smiled wryly. 'It would require blunt force of some significance to create, I'm thinking an implement made of metal or hardwood. It's a complete compound basilar fracture. The bone has splintered. It would have caused catastrophic brain injury,' Ted said. 'Which is what would have killed him.' The human head was able to withstand a phenomenal amount of pressure, but if it impacted the brain then it was game over. Ted reckoned the young man would have been unconscious within five minutes and dead in perhaps twenty, as the arbiter of life swelled and then gave up.

Henry nodded.

'What are these?' Ted asked, pointing to an array of non-organic items.

'Found under the body. A button, a zip, and a synthetic clothes label. Interesting, eh? And remnants of nylon socks, I'd say, bright yellow in their prime.'

Anything man-made was likely to survive the ravages of time and it was always a stroke of luck to find anything that the deceased might have been connected to in life.

'I've checked. The zip has a serial number, and the label has a manufacturer.'

Henry was pleased with himself, and well he should be.

Ted's phone, which he always left on the steel work surface beside his computer when he was working, buzzed and he saw that it was from his secretary. He was expecting the arrival of the body from Parkie's holiday complex outside Keswick. He'd already spoken to Kelly about it, and read the initial notes from the forensic officers at the scene. Apparently, the body was on its way down. There was already a gurney laid out specially for the adult male who'd, it was suspected, died from a head injury and been dragged under a cabin to meet his maker.

Two serious traumatic head injuries in one morning was something that Ted rarely encountered. One old and one new. It was his task to pull himself back into the present and leave Henry to finish off the work on Jack Marrow and concentrate on the man who'd been alive only yesterday.

The job of examining Jack was more academic, in the sense that they had little to go on and Henry's study of the bones, with the help of a zoom call with the osteologist Ted had recommended, wasn't going to get them any more answers as to how long he'd been at the bottom of the lake. Not today anyway. Meanwhile, he'd promised Kelly that he'd get started on the unfortunate man being wheeled down to the mortuary right now. He was a full corpse and had flesh on his bones, and that was more Ted's department. He'd dedicate the rest of the day to it.

Working on the dead wasn't the same as surgery on the living. There was no chance of the patient dying, and so having two examinations ongoing in the same lab was standard. That's if the job had call for it. Ted only occasionally experienced such need, and he was used to taking his time with one unfortunate visitor to his table at a time. The helicopter crash last winter had been an exception, with twenty victims being studied at any one time. He never wanted to repeat that again.

He went to his private dressing room and changed his scrubs. It was important that there was no cross-contamination from Jack Marrow to the current murder investigation. Forensic standards had changed a lot in his time as a pathologist. He could still remember professors smoking on the job and washing their hands only to take a sip of tea.

Technology had transformed their practice, and now everything was labelled and kept sterile until it was seen under a microscope or tested by third parties elsewhere. He had to shift his mindset too. A cadaver not dead for perhaps twenty-four hours required a different approach than a skeleton found in a grave of lake bed mud and dust. He took some deep breaths and got his brain into gear. Kelly had told him they hadn't released the name to the next of kin yet because they wanted a formal ID, but she was sure that it was an old school pal of hers. Ted appreciated the gravity of the situation. Knowing a victim wasn't easy and made all work connected to the body more difficult.

By the time Ted had scrubbed up for a second time and placed his magnified glasses on the end of his nose, the man Kelly called Jason had been wheeled into the mortuary cold-room storage facility. The sound of the gurney stopping and being locked into place, as well as the reflection of the steel on Ted's glasses, was a moment he associated with the true passing of a victim. It was his job to pass on to Kelly everything he could find out about this man, in life and death.

She was due to meet him here at three p.m.

Chapter 15

The only relative of Jason Cooper the police had located to identify his body was his sister, who lived in Penrith, a stone's throw away from where Kelly had gone to school with him and Michelle.

Their parents were dead, and Jason had no other siblings.

Kelly met Joanne Cooper at the Penrith and Lakes, in the cafe. She looked like a woman who'd been inconvenienced. She was much older than Jason and wore the look of a life lived grudgingly all over her face. Kelly greeted her briefly and they walked to the lift, where it descended to the basement of the hospital, to the mortuary. During a heatwave, it might be considered a treat to enter such a cold space, away from the soaring temperatures outside, but not today, under these circumstances. Kelly tried to be pleasant.

'When was the last time you saw your brother?' she asked.

'Dunno. Ten years maybe. He's always been a pain in the arse. I really don't need this but the woman on the phone said they couldn't find anyone else to do it.'

Kelly nodded, trying to think of what else to say.

'What about his ex-wife?'

'Bitch. Never should've married her. I'm not surprised she couldn't get off her arse to come down here. She'd rather ruin my day. Typical. Nothing in it for her, you see.'

'How old are his children? Do you see your nephews?'

'Nah, I'm not keen on kids. They're noisy, and messy. It was bad enough raising Jason on my own, while mam got smashed on lager and I kept Dad from pulverising her. I'm not painting a pretty picture am I? Well, it wasn't.'

'Did Jason not try to get in touch with you to make amends?'

Kelly was momentarily shamed by the fact that she hadn't seen her sister in over a year. Families were unfathomable. Once they broke down, it was almost impossible to unpick the lies and pain to get to some kind of truce.

'He called me when he needed something. I stopped answering. Michelle, bless her, always put up with him. God knows why. She was his best pal.'

It was a display of emotion that was short-lived, but Kelly appreciated the sentiment.

'In here,' Kelly said, stopping at the mortuary entrance. 'It's cold in there,' she added.

'Well it's either him or it isn't. I'm not intending to stay longer than I have to. I'd know my brother's face if he was sixty years old. I'll give you your answer then I'm out of here. I've got things to do. Days don't stop just because somebody's dead.' Joanne grunted. 'Even when he's gone, he's causing me trouble.'

The words cut Kelly, and she wasn't even related to the woman. It was tragic to witness. Surely Jason wasn't that bad? She remembered a fit and energetic individual who carried the crowd. Jason was a born leader, but, from what she was hearing, that had amounted to nothing. He got into trouble all the time, which was part of his allure, but he was never malicious or evil, as his sister was trying to make out. But then she hadn't known him in later life. People changed.

Ted was waiting. He greeted the woman warmly but she brushed him off. Joanne carried a lot of anger and resentment and was constantly on the lookout for somebody to blame. It was a characteristic that Kelly had little time for. Ted bristled too. He got straight down to business and told the woman what to expect. Joanne shrugged off the prospect of seeing a dead body – one that she was related to by blood – with nonchalance.

'Can we get on with it? If I get back to my car in twenty minutes, I won't have to pay,' Joanne said.

Ted led them into the cold-room and opened one of the storage chutes. The sound of the metal drawer opening grated on Kelly's nerves.

Ted pulled down the sheet and exposed Jason's head.

'It's him,' said Joanne. She turned and went to the door.

They had their positive ID.

'I need you to sign some paperwork,' Ted said to her.

Joanne sighed and looked at her watch again.

'It's just here, it'll only take a moment, then you're free to leave.'

Joanne scribbled on the clipboard where Ted indicated, and then left the way she'd been brought in, the door flapping closed behind her, leaving Ted and Kelly speechless.

'What a bitch,' Kelly said.

'Grieving relatives are always surprising,' Ted said. 'Are you coming in? Now that the more unsavoury part of this unfortunate man's passing is dealt with, I can continue the autopsy without her breathing down my neck. Who is she?'

'His sister. Estranged by the sound of it. Can I come and observe for a bit, then I'll need to get back to the office and write up my interviews for the day. I need to brief the team,' she said.

'Come on, let me get you some Vicks, I know how you are in the mortuary.'

She smiled and followed him into his dressing area, where he gave her a mask and a head covering. Then he took her into the mortuary, where Jason's body was being wheeled in from cold storage. He was placed next to Jack Marrow. Henry greeted her.

'Ah, Kelly, thank you for your hospitality last night. What a lovely flat Johnny has.'

His joviality lifted her morbid mood.

'I'm glad you were both comfortable. You're welcome to come and stay anytime, it's much closer in Pooley Bridge than

Dad's place in Keswick, though it is beautiful too. How long are you planning to stay?'

'Well that depends on our man here,' he said picking up a bone. 'It's fascinating. I can base a whole paper on what I've seen already. I'll need to follow up lab results, obviously, and my business is the bones themselves, not the murder investigation, so I imagine I'll be here as long as it takes.'

'We were just discussing cause of death actually, Kelly,' Ted said.

'Oh?' she asked.

'Look,' Henry beckoned her. 'There's a huge trauma fracture – compound and complete we think – to the base of his skull. He was killed by having his head bashed in.'

Kelly shivered slightly, and looked at Ted.

'Any idea what kind of weapon could do that?' she peered at the gaping hole and the way the bone had caved in.

'We think metal – something like a spade or hammer or wrench are all possibilities at this stage. It's my intention to bring a selection of implements into the lab and try each one for size.'

'A spade?' Kelly asked.

'Why?'

'Oh, I'm no expert, I'll leave that to you,' she said. 'How do you know this wasn't done to the skull recently, like by rock movement or something?' she asked. It was important to her to start to establish a timeline. That's what the experts were good at. It meant that when they eventually nailed a suspect, and had him in custody, they could throw the science at him to try to get a confession.

'Well,' Henry cleared his throat.

Kelly glanced at her dad and perceived a diminutive smile at the corners of his mouth. Henry was enjoying this. But she forgave him. It wasn't emotional; purely academic. She appreciated his enthusiasm.

'When flesh decomposes, the bone leaves behind changes, even underwater. If it had been cracked or rolled over – let's say by a rock – then the splinter marks would be clean and sharp. Imagine snapping a biscuit. When the bone is still covered in flesh, and breaks under the enormous pressure of trauma – blunt or sharp force – then the site of the break is wet, messy. It still adheres to the tissue around it and breaks up more like a watermelon hitting the ground.' He smiled at her.

She smiled back. 'Excellent. Any news from the dental lab?' she asked.

'Not yet. But Henry made some interesting discoveries underneath the body,' Ted said. He showed her the button, zip, remnants of yellow nylon, and the synthetic label; meanwhile, Henry beamed behind him.

'Great, they'll certainly help,' she said.

'They'll be sent for analysis as soon as Henry's happy there's nothing left in the dirt for us to find.'

He told her about the serial number on the zip and the manufacturer's name on the label.

'Never heard of them.'

'It could help date him,' Henry said.

'Shall we?' Ted asked her.

Kelly nodded and thanked Henry for his graphic explanations.

Despite having seen Jason unzipped once already, the noise had the same effect as it always did. The buzzing of metal on metal, teeth perfectly aligned and uniform, unbiting themselves in a rhythmic motion, made her arm hairs stand on end.

Two lab technicians were on hand to follow Ted's instructions, and she'd witnessed enough of his autopsies to appreciate his working methods. She pulled up a steel stool and got comfortable, allowing him time to get to know the victim, from walking around the body countless times, to collecting tiny samples with his fine tweezers, to the point where she knew she'd leave: full evisceration.

'The bruise on his face was from a fight yesterday,' Kelly said, as Ted studied it.

'It looks ante mortem. Any other injury from the fight that we know of?'

'No. He was fine after the scuffle and went to the pub. But the other damage, I'm thinking it was when he fell forward onto the stone steps outside his cabin.' Kelly pointed at Jason's face.

'That makes sense, this is ante mortem too. So he was jumped from behind and fell.'

Kelly watched her father pace around the body. She'd dropped Kate back off at Eden House before she came here, to prepare the team for the afternoon brief. From a dip in cases over the last couple of weeks, they now had two homicides – both a mystery – in a very short space of time. It was like that with death. It was never neat and tidy.

Chapter 16

'Been out collecting bodies?'

It was Rob's idea of a joke. Gallows humour kept the team sane.

Kate grinned. 'Christ, they're coming thick and fast. Kelly's at the mortuary, she's coming back to give a brief about five. Where are Emma and Dan?'

'On a late lunch, they were only going for a takeaway, they should be back soon.'

Rob went back to the file he was working on. He was the team's admin cruncher: he lived and breathed details. Kate was still leaning over his chair when Dan and Emma came back in together, discussing sandwich fillings. She noticed the frisson of admiration between them, which was commonly acknowledged as one of the worst kept secrets in the office. But their chat was also about work, and Rob leant back on his chair to stretch, joining their conversation. They were tight, and that was only ever a good thing in police work.

It was Emma's job to retrieve the files of missing males in the area from the last three decades who seemed to match the profile of Jack Marrow. From seven of interest, the list had expanded to eight. They put them in chronological order, ready for when Kelly got back. The photos of the vanished were on the front of each folder, and several of them were black and white images, pointing poignantly to a bygone era.

As she set them out, Kate glanced inside and flicked through the investigations. Police work had remained essentially the same, but methods and technology had transformed it and she didn't recognise the sense in the brevity of some of what she read. She felt a little embarrassed. Some of the files looked as though they'd been simply left to rot. She thought of each family of the disappeared, and felt their anguish. Each of the eight cases had a CLOSED stamp emblazoned over the front of the file. Occasionally, they'd have time to address cold cases, but these had been forgotten in time.

DC Emma Hide had spent the rest of her morning dealing with the press, and chasing the dental records office. She was a dogged operator and had a knack for talking to professionals who didn't want to cough up information prematurely. It wasn't that she rushed people, she just got the best out of them, and they made promises to her that they often came good on. Like now. Emma took a call from downstairs and caught everybody's attention.

'Hey, it's the dental records office.'

Everybody stopped what they were doing and either swung in close on wheelie chairs, or, as in Dan's case, plonked himself on a desk. They all waited and didn't say a word. The only sound was Dan munching on his sandwich from the deli. Kate threw him a look and he shrugged his shoulders as if to say, 'I'm starving'.

Emma replaced the phone and thrust her fist back to her body in triumph.

'We have a name for Jack Marrow. He popped up straight away,' she said.

'Well bloody tell us so we can put the rest of these files away!' Kate demanded.

'If he's there,' Rob said, reminding them that the match might not necessarily be somebody who'd been reported missing in Cumbria.

'Brian Miller,' Emma said.

Kate quickly scanned the eight files. 'He's here. Seventeen-year-old lad reported missing in 1997. His parents were still alive two years ago according to the last update, and living in Penrith.

Bingo,' Kate said. She flicked over the rest of the file but closed it abruptly.

'What is it, boss?' Dan asked in his broad Glaswegian accent in between mouthfuls.

'Look, the boss will want to go through everything as soon as she gets back. Let me go over it and I'll get a grip on the original investigation into the disappearance and make some sense of what, if anything, they missed.'

She went to Kelly's office, carrying the file, and closed the door.

'What was that about?' Emma asked.

Dan shrugged. 'Do I still chase the manufacturer of the silver bracelet?' he asked.

'Of course, if they didn't know about it during the original investigation, it could be vital,' Emma said. 'Get his file up,' she added. Rob tapped the keys on his computer and they read the file together, trying to work out what had spooked their senior officer.

'I think this is what caught Kate out,' Rob said.

He pointed to the signature of the stand-in investigating officer, and it was on every page of the file.

'That's John Porter. Kelly's dad.'

Chapter 17

Ted wasn't in the business of reliving the traumatic last moments of murder victims; he believed that the greatest justice he could pay them was to stay professional and focused. Besides, the pain, agony and final moments of anguish had passed. The cadaver was at peace, finally. Becoming emotional wouldn't help him.

He approached Jason's body as a crime scene investigator when he knew a homicide had occurred: the victim was always the last witness to their own death.

He was still bothered by the corpse's sister, though. Her lack of compassion when viewing her brother had been a blow – not personally, of course, but in a general, more sweeping manner. At his age, he'd learnt that family rifts were conflicts that started small, but ended in tremendous rivalries and vicious battles which, after a time, became irreversible.

Kelly had been as dismayed as he was. He glanced up at her over his magnifying glasses from time to time, to make sure she was holding up. Cutting up dead bodies wasn't for everyone, but as he always told his students: the dead didn't mind. But Kelly was a detective, one with a strong stomach, granted, but a layperson nonetheless. He noticed that she received a call, and he got on with his job.

The man they now had confirmed was called Jason had been a muscular fella in life, though he'd read the file and learnt that the man loved his booze, and it was plain to see. His impressive muscle structure was covered in a layer of flab that screamed stress and toxins. Kelly had mentioned to him that they'd been at school together and he knew that for her to sit here and watch this final act of Jason's life wouldn't be easy for her.

He examined Jason's face. The bruise to the jaw had been explained. Apart from that, he had extensive damage to his eyes, easily explained by traumatic brain injury, caused by the blow to the back of the head. He concentrated on the damage to his right maxilla and made the appropriate measurements and soon concluded that it could be consistent with a fall forward, as Kelly described. He had no other injuries to the anterior of his body. His arms were unscathed, and he could find no defensive wounds. It confirmed that Jason was in all likelihood surprised from behind. His legs were unscathed, so the blow to his head had incapacitated him fully; he hadn't been felled by lower body injuries.

He heard Henry chattering away into his mic across the mortuary, and Ted smiled. The passion that the man held for his job was admirable and one of the things that inspired Ted to pursue his own field. Neither of them could have imagined, fifty years ago, sat listening to a university lecturer droning on about flesh and bones, that they'd end up here, side by side, examining the dead, still trying to get answers for the living, because they loved what they did. Henry's enthusiasm was infectious, and Ted decided that he liked the hum of noise in the room. Sometimes they played music while the technicians worked. Occasionally, they had visitors such as the police, and even barristers in some cases. But ordinarily he was alone, with his silent technicians, and left with his thoughts.

He took a pair of scissors and cut Jason's T-shirt down the middle, leaving his arms inside it like a light summer jacket. In here it was easy to forget the weather outside. Jason's body temperature this morning, when the first forensic crime scene technician had arrived, had been twenty-three degrees. A dead body was like an unfinished cup of tea, it cooled slowly to the ambient temperature of the air around it. Kelly had told him that Jason left work at four p.m. and drank a couple of pints, leaving the pub at six p.m. The human body after expiration cooled at a rate of about one to two degrees per hour; however, given that the air temperature yesterday was over thirty degrees at four o'clock, and had remained high until around nine o'clock at night, when it began to plummet to a low of twenty-one degrees around four this morning, Jason's body temperature would have taken around eight to ten hours to fall enough to match the environment.

They were looking at a window of six p.m. last night, to no later than ten p.m. It was rare to get such a narrow estimate and Ted was pleased that this was something his daughter could definitely work with.

Ted had worked on bodies that had been decaying for much longer than Jason's and now, across the lab, he had one that was a mere pile of bones. There was only one thing sure in life, and it happened to everyone: the death of the human biome and the way it eventually reduced to dust. The only things that changed were the conditions and rate at which it happened.

Jason's chest was smooth and tattooed, like his arms. Ted pressed his hand to palpate over his liver, which, to Ted, looked enlarged, even from the outside. Bloat hadn't yet begun, so the protrusion could only mean one thing: the man was an alcoholic. He had all the signs. Maybe this was why his sister disowned him so readily. Ted knew from his experience with Mary, his ex-wife, that living with an addict was impossibly destructive.

He continued to perform the basic exams of an autopsy, when the body is a crime scene in itself, scraping for particles, deposits and foreign articles under nails, through hair, inside the mouth, inside the belly button and around the genitals, which were then all bagged and labelled, before asking for the body to be turned.

The posterior was far more gruesome. Jason Cooper was a large, once athletic, man and it took three technicians to flip him over. His flesh settled with a wet slap resonant of an exhausted trout being landed on top of a bait box. Ted cut away the remainder of his clothes and took samples from the man's exposed orifices. His boots were removed at this time and Ted peered at them under a light. There was blood spatter on both, to the rear, and he had them photographed, bagged and tagged. The soil around the body had been preserved and sent to an independent serology expert.

Then he moved to the skull. Having handled one bashed-in cranium already this morning, it felt odd to Ted to be repeating the process, but this time at least it was with a specimen that was covered in flesh, with a face attached to it, revealing what the man looked like in life. The injuries were remarkably similar to Jack Marrow's, but Ted tried to shut this out of his mind. It was an occupational hazard of working on more than one body at a time. He had to refocus.

No wonder the man's eyes were so badly damaged, he assessed. The damage to the back of the head was catastrophic. Not only was the cranium caved in, indicating a powerful blow to the bone itself, Ted could see that the man had suffered a rapidly occurring acute subdural hematoma. It would have caused unconsciousness fairly quickly, Ted believed. Not only that, as if for good measure, further down, underneath the head, between the cervical vertebrae, C5 and C7, a huge gaping horizontal wound sliced across the man's neck, above the shoulders, exposing the spine, tearing several layers of muscle, and more than likely, in his opinion, severing the spinal cord. He could see white bone as he used a probe to examine the wound. The gash appeared clean, as if one almighty thrust of a single sharp edge had caused it.

Deep inside, Ted found evidence of uninvited guests. Tiny larvae had been laid deep in the exposed flesh. Their size, inertia and colour indicated that they were freshly laid by the first visitors to the crime scene: the common blow fly, and perhaps its pal, the flesh fly. Calliphoridae and sarcophagi detect corpses from miles away, the Great White sharks of the air, and find them with the precision of a NATO weapon guidance system. In heat like they were experiencing in the Lake District this summer, the first hungry winged diners would have arrived within hours of Jason's last breath, seeking out the sweetest offerings at the wound sites.

He picked up a pair of fine tweezers and retrieved some samples for later study, but he knew from experience, and working closely with etymologists, that they were young. He also photographed deep inside the laceration and directed a technician to add to a computerised diagram of injuries on his computer.

Kelly returned from her phone call just as he was lifting a tiny fly egg out of the cavity.

'Nice,' she said.

'Are you all right?' he asked. 'You look pale.'

'I have some news.' She got Henry's attention, who dragged himself away from the pile of bones that was growing. He'd recovered seven ribs and fragments from the soil and added them to the gurney acting as Jack Marrow's final resting place.

'I've got a name. Dental records came back to us. You were right. He's been down there for twenty-four years. He disappeared in 1997. His name is Brian Miller.'

Chapter 18

Kelly was shaken and she desperately needed a coffee, or even something stronger, but that would have to wait. She'd only had the stomach to tell her father the identity of the pile of bones. After she'd done so, she'd left the mortuary in a hurry, telling him, and Henry, that she had to take another urgent call.

She hadn't even stayed to listen to Ted's discoveries so far from the continuing autopsy of Jason Cooper.

Henry and Ted hadn't noticed her nerves. They'd toasted the passing of Captain Jack Marrow, now they could call him by his real name.

The sinking feeling she'd had all morning came back. It had started when she first saw Michelle. It was as if memories, tangled in the past, crept up her spine, but she couldn't pinpoint why it was so disturbing. Then seeing Jason, and being reminded of all those years ago when they were barely young adults embarking on their futures. Maybe it was because, for one of them, that was never to be.

The news of Brian's disappearance had gripped their community at the time.

Kelly knew his family. That was the thing about going to school in Penrith. Everybody knew everybody. That was part of the reason she'd run away to London, to escape the suffocating incestuous cloak of her past.

Kate had told her that the file on the case of Brian Miller was waiting for her in her office, but she didn't need to see the file to remember Brian's face. The nagging siren clawing at her memory that she'd felt all day yesterday was now explained, but she'd never for one second imagined that it would be Brian down there. Even after twenty-four years, his clothes, his cologne, his swagger and his walk all remained seared into her memory, shut off, but now fully reanimated. Brian had been one of the popular crew, and, as such, considered a problem. He was of the same crowd as Michelle and Jason.

The cool group.

At school, they caused teachers no end of headaches with their antisocial behaviour and back chat. Their disregard of the rules – or any kind of conformity – landed them in detentions almost daily. Back in 1997, before the prevalence of mobile phones and social media, their power was wielded by mouth and reputation. Kelly yearned to be like them, but she never was.

For some reason they left Kelly alone, though she sat with Michelle and Jason in geography. She'd even helped them with their work, or rather, they'd copied her answers. But they didn't talk much. Even a glance from Jason was enough to send Kelly into a silent spin of shame and runaway thoughts that her heart couldn't take. She fantasised about him, and how he would one day see her as a real person, despite her lack of conversation and her cheeks turning pink if he so much as looked at her.

Michelle noticed. So did Mr Thompson, their geography teacher.

Their group smoked behind the trees at the end of the football pitch, and their school uniform smelled of sweet weed after break. They invited each other to secret parties and it was rumoured that the girls in the group lost their virginity at thirteen years old. They didn't care. The worse the stories about them, the bigger their egos got.

None of them wanted to attend sixth form college but there was nothing else to do. It was just another excuse to avoid the world of work for another two years, and stick together. But by the time the term began in September the group had mysteriously broken up. Michelle and Jason spent all their time alone, and Brian drifted off to find other friends. There were others in the rift. Dave Crawley was one, and Kelly's skin crawled when she thought about him.

When Dave asked her out, towards the end of sixth form, she'd accepted. He was everything she wasn't: confident, risk-taking, and outgoing. It had been a mistake. In the end, the only way to get away from him was to run away altogether. No one ever understood why. Her taking the blame for breaking his heart, for being aloof and snooty, was a small price to pay for her freedom. Toxic people love to hide their secrets behind a scapegoat and she didn't mind. Her skin had grown thick in London, and when she arrested him for trafficking sex workers back in 2018, it was all worth it, to be finally vindicated.

She wondered, now, what had happened to the others. After twenty-four years, she was about to make it her business to find out.

Brian didn't make it through sixth form, he disappeared before the end of the first year. And Michelle and Jason dropped out in the second. She recalled many of the feelings she'd had at the time, as well as the shock felt by the local community. Her peers lived the horror of never knowing what happened to him, but also because her dad was closely involved in the case itself. She remembered him coming home late, shaking his head, warning her to stay away from 'that crowd'. The whispered conversations between him and her mother. The secrecy. The shame and guilt at never finding him.

The case had bothered him. She'd heard him talking to her mother about it, late at night, when they thought their daughters were both asleep. He never forgave himself for not solving the case. John Porter wasn't a detective, but in those days, small departments like Penrith, faced with major cases like a missing seventeen-year-old, pooled their resources where they could. John, as a uniformed officer, did much of the legwork that formed the backbone of the inquiry. Only in later years did Kelly come to appreciate how hard he worked.

He died not knowing what happened to Brian Miller.

Chapter 19

Kelly's office at Eden House was sparse, functional and too damned hot. She sensed the subtle change in mood as soon as she walked in, and made her way straight to her private office and closed the door. It didn't take Kate long to knock on it, come in, and plonk herself down on a chair. She'd told Kate about her link to the victim.

'I put the file on your desk so you could go through it before you decide how much of the original paperwork to explain,' Kate said.

So that's what this was about.

'Thanks, I appreciate that, but, you know, it's absolutely fine. If mistakes were made I'll find them and deal with them. John Porter was a good operator, by reputation. My biggest fear is finding out that he wasn't as good as everybody remembers.'

'Do you want me to have a word with the team?' Kate said.

'And say what? If anything needs addressing then I'll do it. But thanks, I appreciate the thoughtfulness.' Kelly sighed and wafted her face with a file.

'He wouldn't have had the resources available that you do now. And he didn't have a body.'

'I remember everybody saying that Brian had just wandered off, to find himself, to make his fortune, to escape the arse end of nowhere, to find his freedom. And all this time he was at the bottom of a lake, his skull caved in. By who? Did his killer have more victims? And am I the right person for this? Is that what you meant by have a word? Maybe I shouldn't be anywhere near this case. My father investigated it, the victim was a school pal.'

Perhaps it was the heat, or lack of sleep, but Kelly was irked.

'Rubbish, get a grip, woman. You're the only one who should be reopening this case. And I'll run naked all the way to the castle if you don't catch the bastard who did this. Kelly, straighten your crown and get to work. Self-pity stinks of shit on you.'

Kelly looked up at her friend and smiled.

'Right, ready?' Kate asked.

They walked into the incident room at five p.m. and greeted the gathered team. The superintendent had approved extra pairs of hands for the two cases, given the bodies were found within a day of each other, and the workload would be enormous. But, for now, she briefed her crime unit team alone. From there, they'd each take responsibility for an area of the two cases, and delegate to uniforms loaned to them from downstairs, as well as surrounding departments, at their own will.

'Well, it's been quite a couple of days,' she began. 'I need to clear up something for you all now. I knew both of the victims. I went to school with them. It is personal. I'm not a robot. However, I give you my word that I'm throwing myself into this like any other investigation we've worked on as a team over the past few years. I'm applying the same exacting standards to my own inquiries as I always ask from you. I believe my knowledge of the victims will benefit our inquiries, and as always, I'll be expecting full transparency. This is a small town. Some people that we come across while picking apart the lives of Jason Cooper and Brian Miller might have a thing or two to say about me. Believe me, I can handle it. There may be animosity towards my father, John Porter, who was closely involved in Brian's case. Again, I'll handle it. Just to clarify, John Porter was the father I knew for almost forty years. It only came to my knowledge recently that my biological dad is Ted Wallis. And that's the first and last time I need to say anything on that. It's straight to business.'

Her team responded in the best way she could have hoped, by saying nothing, but showing that they couldn't wait to get started. They looked at her, not with shock or immature fascination about her past, and how it might come up in their current investigations, but as they always did: waiting for her lead.

It was a relief. Kate winked at her.

She dealt with Jason's case first. They had little to go on yet. They had a body, they had evidence of homicide by blunt force trauma to the head, causing catastrophic brain injury, and they had time on their side. There was the locket, knowledge of, she stressed, should be kept within the team. But that was it. Jason had no family to speak of, excepting his estranged wife and kids, though Kelly wanted the sister checked out: Joanne's lack of emotion was a red flag to her. She wanted to chase any financial arrangements Jason might have had with his ex-wife, and they also had to check out Danilo's alibi at the farm and Jason's alleged sideline business in farm tools. She'd requested Ted rush through his toxicology, because she suspected that he was probably inebriated when he left The Swan Inn, and Connie was covering for him. She also wanted a more detailed statement from the person who seemed to be Jason's only ally in this world: Michelle Parkinson.

As far as Brian Miller was concerned, the team's first task was to familiarise themselves with every minute detail of the original missing person inquiry. Then they'd eagerly await Dr Henry Dempsy's findings.

They had a punishing evening ahead, and the air-conditioning unit had just packed up.

Chapter 20

At gone nine o'clock, Kelly kicked off her trainers at the door and prepared herself for the hundred-mile-an-hour incoming Exocet missile that was her daughter. Lizzie raced towards her in her walker and crashed into cupboards and doorways on her way. Kelly sat on the floor to receive her and kept an eye on her feet, in case they be sliced off by the speed of the passing plastic undercarriage of Lizzie's artillery.

'Hi Kelly,' Josie said brightly as she came out of the kitchen.

'What's that divine smell?' Kelly asked.

'Churros. In the oven.'

Kelly wished she could recapture the excitement of youth over such simple treats but all she wanted was a cup of tea, followed by a large glass of red, not necessarily in that order.

'Where's your dad?'

'Oh, he was called out to an accident at Wastwater, somebody fell on the coastal path, under the screes.'

'That sounds nasty. I wouldn't walk that path if you paid me,' Kelly said. The walks above the screes on the lake's east side were truly spectacular, and Kelly wondered at the desire to climb over loose rocks with boulders the size of houses waiting to fall and crush anything in their path at any moment, when the views were so much better from the top. The vision made her shiver, though all the windows were open and it was still twenty-six degrees outside. They'd walked there a few times since the terrible helicopter crash last winter. On a sunny day, Illgill Head had unparalleled views from the top, with Scafell Pike to the north and the grand dome of Great Gable at the head of the lake. From the summit, Wastwater looked deep purple and serene, like coloured glass.

She retrieved Lizzie out of her walker and her daughter demanded she walk her around, holding her hands above her head. It was her favourite game at the moment. Kelly knew she'd have to do at least ten circuits of the ground floor, before Lizzie sought something else to occupy her. She was getting strong. The mindless repetition of the pacing emptied Kelly's head and she breathed easily for the first time all day. The late evening sun threw shades of gold and bronze over the terrace out the back. They'd had a safety gate fitted across the double doors, as well as across the stairs. Their house had changed to accommodate the little powerhouse in a frilly dress. Kelly always knew when Josie, not Johnny, had dressed her. Johnny chose sturdy practical gear, as if he were readying her for a hike. Josie picked out cute pieces, with matching cardigans and socks, and tried to put pigtails in her fine hair. Lizzie usually pulled them out.

The whole way around the third circuit, Lizzie chuntered to herself and to her mother and Kelly answered her as if she could understand.

'Really? That's fantastic! I agree!'

It encouraged her daughter more and by the time Kelly took a sip of wine brought to her by Josie – skipping the tea – it would seem to an outsider that they were having a full-blown conversation about world peace.

Images of bones, flesh, spades and gallons of water pressing down on corpses fell away and Kelly began to relax. In the presence of so much life, she couldn't help it. She led her daughter to a puzzle on the floor and Josie came in with a plate of churros. She offered one to Kelly, and, against her better judgement, she took one. It was warm, sweet and soft on the inside. Chocolate sauce squirted out of the side of her mouth and Lizzie clamoured for some. She allowed her daughter to suck some of the chocolate from the treat and she squealed with delight. Government guidelines forbade sugar for the first year of development but Kelly figured that if you denied a child a treat, you made it an obsession, and she knew what happened when kids became fixated on something they couldn't have.

She briefly considered what caused children like Lizzie to turn into monsters and anxiety gripped her. Didn't even serial killers have parents who, at one time, loved them? Weren't the most depraved torturers once suckled by their mothers? She pushed the thoughts away. Motherhood had brought with it a tsunami of guilt and reminders of mortality.

'Did you look at those rooms online?' she asked Josie about her halls of residence in Durham, to change the mood inside her head.

'Yes, they're so nice, but really expensive,' Josie said. She sat with her legs crossed, tapping her foot, chewing the Portuguese fried donut.

'Your dad and I will help, we want you to be comfortable. Everything is so expensive now. My degree was free.'

'I know, I can't believe it, it's so unfair.'

'I guess it was unsustainable and a lot more kids go to uni now. That's what the loans are for. Take the maximum offered, that's my advice. You can always pay it back. You'll have the time of your life and you shouldn't be worried about money.'

'Thanks, Kelly. My mum said dad should pick up the bill because he's retired and can clearly afford it, but that's not the point really, is it? Look at you, you're not even my mum and you're so concerned over me just enjoying the experience, rather than bickering about who should pay for it.'

'Did you speak to her last night? I heard shouting.'

Josie's phone calls with her mother – Johnny's ex-wife – never seemed to go well.

'Yeah. I put the phone down on her.'

'Oh. I'm sure she means well.'

'You always say that. I'm not a kid anymore, I know when somebody is using me to get to the other one. Dad never does that, and neither do you.'

'She's hurting. People always get angry when they're in pain.'

'Oh, Kelly, you always have an answer for everybody's behaviour, don't you. Why can't you just admit that she's a bitch? I did a long time ago. I'm an adult now. Why is it that people always say "but they're family" as if that makes it all okay?'

Kelly smiled. She'd always tried not to get involved in the relationship between Johnny and his ex-wife, though Josie's stern assessment was one she'd agreed with for a long time. Johnny hadn't been an easy partner, she knew that. And not the greatest dad either. He'd put the army first, and been selfish and immature. But that was then. He'd repaired the relationship with his daughter, this year more than ever. They spent time hiking together, and sailing the *Wendy*, coming back laughing and telling Kelly what they'd done. They took Lizzie out and spent time with Ted when Kelly was crazy busy.

They'd gone shopping for bedding and pots and pans for Josie's new accommodation.

Josie's departure would be a wrench for Johnny, Kelly knew that. She prepared herself for his sorrow when she finally walked out of the door, and she tried to buoy him by reassuring him they'd go and visit. Durham was a beautiful city, they could spend a few nights there, get a babysitter for Lizzie. Go and visit the Light Infantry memorial at the Cathedral. See the resting place of the Venerable Bede.

Perhaps even have a lie in… make passionate love like they weren't parents, get drunk and fool around.

'What are you smiling at?' Josie pulled her out of her daydream.

'I'm jealous of you starting your life. And excited. Lizzie won't know where her playmate has gone.'

Josie finished her food and grabbed Lizzie, swinging her around and making her giggle. Kelly flopped onto the sofa and finished her wine.

The front door opened and Johnny walked in. He looked tired, but buzzing from the fresh air and exercise, even though a rescue wasn't a leisure pursuit.

'Everything go okay?' Kelly said.

He went to her and kissed her. He was hot from his exertions, and his face was tanned. He smelled of the mountain.

'Yep, one casualty, airlifted to the Penrith and Lakes, they're doing fine. I think there should be warning signs on that path.'

He sat down next to her on the sofa and Lizzie fell to her bottom and crawled to him.

'How was your day?' he asked.

'Oh, I've just been telling Lizzie all about it.'

Their daughter cooed as if in agreement.

'That bad?'

'Two casualties, not airlifted, not okay.'

'Two? I thought it was a pile of bones?'

'Another one came in. This time fresh as a daisy, hours old. And we got names. I knew both of them. They were old school pals.'

'Really?'

'God, you two are weird. I know what you're talking about. And one of these days so will Lizzie. You'll have to change your little chats then. Missing the odd word out here and there really doesn't cut it, I know exactly what you're talking about. I heard about the bones on the TV, and Granddad told me. What does the skeleton look like, is it a ritual sacrifice like everyone is saying?' Josie asked.

Kelly glanced at Johnny.

'No. He was a young man, your age. He went missing twenty-four years ago.'

'Amazing! What colour are the bones? Are they still white?' Josie asked.

Kelly was used to Josie's inquisitive mind and she didn't think the conversation morbid or strange. She shook her head.

'More mustard.'

'Cool. And how did you get a name?'

'His teeth. They matched our dental database. We knew he couldn't have been down there for much longer than a few decades because he had metal fillings.'

'Why did he have metal fillings?' she asked.

Kelly chuckled. 'Because that's what they used to shove in your mouth, the white ones hadn't been invented yet.'

'Look,' Johnny said, bending his head back and opening his mouth, showing a collection of amalgam fillings in his molars.

'Dad, that's disgusting.'

Lizzie copied him, and threw her head back.

'Can I bathe her?' Josie said.

Kelly nodded and Josie picked Lizzie up, taking her upstairs.

'You okay?' Johnny asked.

'Should we get the last of the sun?' Kelly suggested. She refilled her glass and poured one for Johnny, and they went outside. The evening glow of the dying rays illuminated the wooden terrace and the heat accumulated all day radiated around the space. The sun beds were warm and they took one each, placing sunglasses on their heads and lying back.

'How well did you know them?'

'I'd forgotten about them both for twenty years. Now, I look back and I'm remembering moments in class, when we laughed. Parties, when I watched them dance. A school trip, after we sat our GCSEs, canoeing on Derwent, them messing around and tipping people out of their boats. Stupid things that were pushed to the back of my mind for an age. It reminds me that I lived a whole life here before I left. I'll have to make contact with everybody they knew. At least one is in prison and I put him there.'

'Dave Crawley? The one you were engaged to?'

She nodded.

'You won't have to see him face to face?'

'It's not necessary, but I kind of want to. He was pretty close to Brian.'

'Brian?'

'Mr Skeleton. The one who we found this morning was Jason. Jason Cooper.'

'And they were both in your year?'

'Not just my year, my classes. They were the cool group.'

'You mean you weren't in the cool group?' Johnny teased her.

She threw him the finger, and lay back to bask in the disappearing sun.

'So, they weren't accidents then?' Johnny asked.

She shook her head. 'They were murdered. Twenty-four years apart, but in a strikingly similar fashion, from what Ted told me today.'

'So, you think they're connected?'

'No, it's crazy to assume that at this stage. But it's disturbing, and it makes what I've got to do a hundred times harder. "Hi, I'm Kelly, remember me? I'm looking into the murder of Brian Miller twenty-four years ago, oh, and by the way, Jason Cooper was bumped off yesterday too. You knew them both."' She mimicked herself and rolled her eyes under her glasses, and took a large gulp of wine.

Johnny got off his lounger and nudged her, squeezing onto hers. She moved up and allowed his body to settle next to hers.

'Well if they are connected, you're the best person to find out how.'

Chapter 21

Kelly checked herself anxiously in the mirror.

She'd slept badly. The red wine hadn't helped. That and her erratic approach to nutrition. At gone ten o'clock, Johnny had decided to get fish and chips from the Inn on the Square. It was damn good but she'd had a churning stomach all night. Now she felt sluggish.

She kept changing her clothes. Today she was driving to the house where Brian Miller had lived all his young life, and his parents still lived. They'd faced twenty-four years of hell, never knowing where their son had gone. It was to be her who got to tell them the news. It was bittersweet. On the one hand, they could now bury Brian, and say goodbye, when the coroner had finished with him. But on the other, he was never coming home. That tiny ray of hope that the parents of the missing harbour for the moment their child might just walk through the door and say, 'Hi Mum, hi Dad!' was now never going to happen.

Johnny came into their bedroom with Lizzie. He'd been up early with her and he looked tired.

'Nervous?' he asked, lying across the bed, watching her, putting Lizzie beside him.

'Yeah. I want to do Brian justice.'

Last night, he'd listened as she told him the stories she remembered. Moments from her past, involving Brian. It was as if she was getting to know him all over again. It was funny how memory changed over time. If she'd been asked a week ago about Brian, she'd have said that he was a typical annoying teenager: arrogant, rude and stupid, but hilariously funny and sometimes incredibly sweet. He got into trouble. He smoked weed. He was disrespectful to teachers. He cracked inappropriate jokes. But now, she focused on the positive recollections; as the class clown, he provided light relief in science when the experiments were mind-numbingly dull, and he exposed himself when they went for their swimming lessons to the Penrith pool, which, at the time, was hilariously entertaining. Nowadays, he'd be hauled off for psychological testing, and likely expelled.

He was an ordinary kid.

But then she remembered that he always found an excuse not to join in the swimming lessons. Brian couldn't swim, and he was too embarrassed to learn. She knew this because he'd fallen into the lake at Derwent, on their trip, and he'd been scared out of his wits. In fact he'd accused a teacher of pushing him in.

She refocused.

'You'll do him justice,' Johnny said.

Maureen and Donald Miller were hard-working ordinary parents. They loved their only son, and had lived without him, not knowing where he was, for twenty-four years. She knew that their agony, which she was about to burst in on, was incomparable in scale and depth. How could she do it justice?

But she had to face them, because, like it or not, they were part of her inquiry. She'd read the file, mostly scribed by John Porter, well into the night. It was another reason she hadn't slept well. Memories of her own family flooded back. It was a small community and the disappearance had rocked it. It was an ugly wound that was about to be reopened, and she was the one with the unenviable task of picking off the scab. Her inquiries wouldn't be welcomed, and neither would her physical presence.

She opted for a black skirt and loose white blouse, with a light black jacket and pumps. She threw her lanyard over the top and turned for Johnny's assessment.

'Sensible, respectful and professional,' he said.

'Okay. I'm ready.'

'You need to eat.'

'I haven't got time.'

'There's a bacon sarnie going free downstairs and it's still warm.'

'Thank you,' she smiled.

He followed her downstairs and she gave Lizzie a cuddle before heading off.

She retrieved her bag from the back of a chair and said goodbye, holding the bacon sandwich in her other hand, closing the door quietly behind her as if that might soften the impact of what she was about to do. The family of Brian Miller had yet to be formally notified of the identity of the skeleton found in Thirlmere.

That was her job.

She pulled away from the house, chomping on the warm bread, wishing she could just stay under her duvet for the day, and headed in the direction of Penrith, but not to her office, to a small estate on the edge of the town, a stone's throw from her old school. She'd driven past it probably a hundred times since returning from her stint in the Met, but she'd never imagined that it would spook her as much as it did today. She wondered how many of her teachers were still in employment there. Some of them might even be dead by now, a couple were virtual corpses when they were teaching her, that was for sure.

Her inquiries would take her to the school in good time. The sixth form college was connected to it. It was essentially the same building but it liked to think of itself as something separate. Something better. The head teacher of both was still Mrs Gooch. Kelly had checked. She must be a hundred years old by now. She'd also checked if Mr Thompson still worked there. He did.

She approached Penrith and almost took the turning to her office out of habit, but checked herself and drove past the school gates, peering at the 1960s concrete structure, expanded and modernised since 1997. It was eight thirty in the morning and kids wearing the same dark blue uniform messed about, hitting each other with bags, scraping their shoes and piling onto crossings when the lollipop lady ushered them across.

She spotted the sixth-formers, because they wore their own clothes, and they looked serious and studious, peering down on the younger years; embodiments of youth struggling with adulthood. The boys looked so young and she remembered Brian the last time she'd seen him. His smile. His dirty jokes.

His leather jacket.

Shit.

She pulled into the side of the road when it was clear and parked, hammering her mobile phone. Kate answered on the third ring.

'Kate, Jesus, Brian had a leather jacket!'

'Morning,' Kate said.

'Sorry, morning.'

'What are you talking about, Kelly, I'm not even awake. I'm on my first coffee. Where are you?'

'I'm outside Westmorland Comprehensive.'

'What?'

'On my way to see the Miller family.'

'Of course, sorry. Have you changed your mind? I can be ready in ten minutes and meet you there?'

'No. I just remembered, Brian wore a leather jacket to sixth form, like, all the time. It was old and brown, and had a thick biker zip. I swear it was in a photo released by the family. It wasn't in his file, but I remember. It must be driving past the school that's jogged my memory.'

'Okay, great, what do you want me to do?'

Kelly laughed. 'I don't know, I guess I just wanted to tell somebody.'

'Are you thinking that the leather to bind him might have been cut off his jacket?' Kate asked.

'That's exactly what I was thinking. If we find the manufac-turer of the jacket – it might even be the label Henry found – then we can see if the fragments match. I'll ask the Millers for the photo.'

'Good luck. You've got this. It's the worst job. Whatever their reaction, you're giving them peace at last. The liaison team will be there at nine thirty.'

'Thanks, Kate. I'll see you when I finish.'

Kelly ended the call and drove away, towards the Millers' house. Her spirits were buoyed by the recollection. The closer she got to this case, the more her memory was kicking into action, and she wondered what else the day would throw up.

Chapter 22

Maureen Miller answered the door.

Kelly smiled and showed her lanyard. Maureen opened the door wider. Kelly reckoned that she was probably only in her sixties but she looked twenty years older than that. Agony does that to a face. They were expecting her and had been told there'd been a development in Brian's case.

She went in and found Donald watching TV. He, too, appeared much older than his years but he managed a weak smile. He turned off the TV. Kelly battled to control her intake of breath when she saw the portrait of their only child hanging above the fireplace. It was a large image of Brian, probably two feet across and longer down. It looked like a painting. Brian beamed down at her, and Kelly swallowed hard, accepting a cup of tea and a comfy chair. In the picture, Brian wore a brown leather jacket. She could see that, on his chest, on the right side, there was an emblem sewn onto it.

'What a lovely picture,' she said.

'Thank you, it's our Brian as we remember him. What did you say your name was?' Maureen asked.

'Detective Inspector Kelly Porter.'

'Are you related to John Porter?' Donald asked.

'I am, he was my dad,' Kelly said. It wasn't worth explaining the ins and outs of her family dynamic. For today's purpose, she was John's lass.

'Well I never. You look like him. Maureen, she looks like him.'

Kelly smiled, finding it comforting in a strange way that they should see a resemblance when it wasn't John's blood coursing through her veins.

'You've taken on where he left off, then,' Donald said.

'I have. It runs in the family,' she said. The connections made the atmosphere less tense.

Maureen put a cup of tea down beside her and she sat down. They both looked at her expectantly.

'What's the emblem?' she asked, taking their attention back to the picture.

'Don't you know? Goodness, lass. Don't you recognise the Kopites?'

'Kopites?' Kelly asked.

'Liverpool. The football club.' Donald chuckled.

It was a natural and warming utterance, penetrating the sombre reason she was here.

'We call ourselves the Kopites, after the Kop, the fan's end that was made all-seating after the Hillsborough tragedy. I took Brian there in 1994, during the last season when you were allowed to stand. There was nothing like it on earth,' Donald explained. 'He sewed it on himself.'

The touching personalisation caught her off guard and made what she was about to tell them even more difficult. All she could do was dive straight in and save them more anguish. 'I've come to inform you both that we've discovered the remains of a young man in Thirlmere reservoir.'

'The skeleton in the lake? It's our Brian?' Donald asked meekly.

Kelly nodded.

Maureen let out a cry and covered her mouth.

'I'm so sorry,' Kelly said. She waited, as Donald got up to comfort his wife. There was nothing else to say until they'd gathered their thoughts. She sat staring at her cup. The sounds of whimpering and whispers filled the room and she peered up at Brian, hoping that she could find his killer after all these years. She almost said 'I'm sorry' again.

'How do you know it's him? DNA and the like?' Donald asked, holding his wife tightly. He was obviously the mouth-piece of the pair. Maureen watched her and brought a tissue out of her cardigan pocket. Kelly reckoned that she kept a daily supply of them in there.

'Dental records.'

Maureen wiped her eyes.

'All these years,' Donald said. 'He's been in there the whole time?'

'Yes, we think so. There is strong evidence to suggest that he has been in the lake since he disappeared.'

'So he fell in? What the hell was he doing over at Thirlmere?' Donald asked.

'Maybe it's not him. He couldn't swim, he wouldn't have gone in a lake,' Maureen said hopefully.

'Dental records are conclusive, Maureen. And the age of the remains matches.'

'Remains? Oh dear God, he's been under there this whole time, been picked at and cold, and all alone. I can't bear it, Donald,' Maureen sobbed.

This was what Kelly was dreading. She felt their pain like a kick in the guts. She couldn't possibly imagine what they were going through right now. After the torture of having a son missing for twenty-four years, now to find out that he'd been at the bottom of a lake, when they'd kept hope at least somewhat alive, as anyone would, was crushing.

'There's something else I need to tell you,' Kelly said.

Donald and Maureen stared at her.

'Why don't you go and slice that cake, love. I'll talk to the detective,' Donald said to his wife. Kelly saw the man's compassion and his love for his wife. He was protecting her. He could tell her later, in his own time, and in his own way, to lessen the blow. She nodded and left the room, closing the kitchen door gently behind her. Kelly heard her let out a gut-wrenching cry.

'Is it bad?' Donald asked simply.

'I'm sorry, Donald, it is, yes. Brian has been exhumed and examined. He was murdered.'

'Dear Lord. Murdered? No. My son.'

Kelly looked at her hands. She wondered how on earth people coped in situations like this and she knew the answer: they didn't. She wondered, too, what her reaction would be if some detective ever knocked on her door and told her that Lizzie had been murdered and dumped in a freezing cold lake. Somehow the dumpsite made the crime worse. Brian's solitude all these broken years was an extra burden for his loved ones. She waited as Donald composed himself to speak.

'Was he examined… respectfully?' he asked.

His eyes had turned puffy and red but Kelly saw no tears. Maybe he'd run out of them years ago, but his body still produced the other symptoms of sorrow.

'Yes, he was. I was there. The doctor who performed the procedure has much experience in these matters. We want to make sure that we find every possible detail that could help us find out what happened.'

'And catch the bastard who did it?'

'And catch the bastard who did it,' she replied.

'How did he die?'

'We think he was hit on the head. It would have been quick, Donald. The pathologist told me that.'

'So he wasn't drowned? I have a fear of drowning. Especially in cold water. I don't go near the lakes. Brian couldn't swim, he'd be terrified.' He repeated his wife's sentiment.

'We can say with authority that he would have been deceased before he was put into the lake,' Kelly said.

'Thank you. It's a blessing.'

Kelly allowed Donald to take a breath.

'Can we bury him?'

'Of course. We need to complete our inquiries, and I will make sure that you get Brian back as soon as possible.'

'What are you doing to him?'

'We want to make sure that all the tests at our disposal are carried out. We don't want to miss anything. We're reopening Brian's case as a murder investigation.'

'But what tests? Are you cutting him up?'

Donald's face broke and his shoulders shook.

'No. We're taking samples from his body, and we've found some items too.'

Donald gathered himself. 'What items?'

'A zip. Was Brian wearing that jacket when he disappeared?' she asked, nodding at the portrait.

'Yes, he was. We told the police – your dad – at the time.'

'Do you have any photos of Brian shortly before the last time you saw him?'

'We do. I've got stacks of them. I can tell you exactly what he was wearing. He had on his blue jeans, a white T-shirt – Bon Jovi – yellow socks, and that jacket. He has a pair of jeans exactly the same upstairs, do you want to see?' Donald asked.

Kelly could hardly say no. She followed him up the narrow staircase to what was Brian's room and Donald showed her inside. It was frozen in time, and Kelly recognised the posters that she too had in her room twenty-four years ago: Oasis, Boyzone, Coolio and the Backstreet Boys. His schoolbooks sat in a pile neatly on his desk. The duvet cover was Star Wars themed and Kelly smiled privately and whispered to Brian that she wouldn't tell anyone. She could smell a faint whiff of cologne mixed with the musty aroma of old clothes. Donald opened the wardrobe and showed Kelly Brian's clothes. He pointed out a pair of jeans and Kelly examined the zipper. It looked identical to the one they'd found.

'Donald, I've got something else to show you,' she said.

She took a plastic evidence bag out of her rucksack and passed it to him.

'Do you recognise it?' she asked.

It contained the silver bracelet that had been found around the skeleton's wrist, or where his wrist would have been in life.

'It's Brian's. It was his last birthday present, where did you find it?'

'With his body.'

Donald fingered the bag and felt the links of the chain underneath the plastic, as if touching it would bring his son back. 'He said he was going to meet friends, and he never came home.'

Kelly had read the file. Brian had said he was going to meet Michelle, Jason, Paul, Dave, Carol and Tracey. The usual crew.

She knew where Dave, Michelle and Jason were. Dave was banged up. Michelle had just found a dead body, and Jason had his head bashed in yesterday. Paul lived in Pooley Bridge, though she never saw him. They'd stopped swapping pleasantries when she put Dave Crawley behind bars. Carol and Tracey lived in Penrith. The gang had never moved away. For all their swagger that they ruled the world, they hadn't got far.

Chapter 23

Kate worked in Kelly's office. She read through the closed case file of Brian Miller. It was her job to reopen the inquiry and start a new report. The findings of Dr Henry Dempsy had yet to be completed, so she began where John Porter had, twenty-four years ago.

Brian's parents had last seen him on the twentieth of July, 1997. It had been an evening of inconsequence. Brian had rushed his tea of beans and chips, eager to get out into the sunshine, his mother said. He'd skipped his homework, as was the norm, to the chagrin of his father. He'd left his room a tip, which was also normal. Kate read what he'd been wearing and knew they had the right body. It wasn't that she'd doubted it, it was just that when faced with a skeleton, like she'd seen stuck in the dusty dried lake bed of Thirlmere on Tuesday morning, it was difficult to associate it with the life that had once coursed through the boy's organs and nerves. She pondered life's fragility. It was an electrical miracle, switched on one minute and then the next, extinguished. She'd seen people die. Her mother. Her grandfather. She'd watched the last breaths of the dying, and she knew Kelly had too. But a pile of bones was different. It was so de-animated, so utterly devoid of life that she was keen to breathe that vitality back into Brian. Getting to know the victim was part of working out a motive for murder. With a motive, they'd be closer to catching the killer, if they were still alive, or even in the country. Twenty-four years was a bloody long time.

She pictured Brian in his leather jacket and white Bon Jovi T-shirt, his blue jeans and his yellow socks. What type of teenager wears bright yellow socks? A confident one. One trying to get attention. It fitted his personality, according to his parents. Brian was funny, cheeky, loving and energetic. That was her starting point. Kate knew that a mother's love was blind, like her own. But what did others think of him? His teachers said he was disruptive and failed to realise his potential. She'd heard that before. He messed around in class. He didn't suit the discipline required for A Level study. His attendance was appalling, and so was his attitude. He was popular, though. And he was reported to have been part of a tight group of friends.

She read their names.

Each had given a statement to the police when they'd been interviewed, in the presence of their parents. They all told the same story. Brian had met them in town, by the castle, and they'd all piled into a car driven by Dave Crawley. All six of them. Jason Cooper couldn't come out that night. She noted that John Porter had given them all a lecture on seatbelts and the horrific consequences of driving without one.

They'd driven to Keswick and parked in the car park by the lake, picking up some alcohol on the way. A witness had corroborated this. They'd taken their cans and bottles to Friar's Crag, and sat around watching the sun dip behind Catbells, drinking and smoking.

Another witness spotted the youths and reported them for anti-social behaviour.

They'd moved on, piling back into the car around ten p.m. All the witness statements from the kids matched up. Dave had driven them back to Penrith and dropped them in town, then he'd gone home. That left five, who all lived close by. From the centre of town, Paul's house was first, but he'd insisted on walking the girls home. Carol and Tracey had to be helped into their houses by their parents. Michelle, Brian and Paul then said their goodbyes, with the boys watching Michelle go into her house.

All the parents corroborated their stories, except Brian.

He never came home that night. But the others did.

The corroborated statements eliminated them from the police inquiries at the time, but now, the question remained, could any of them have got to Thirlmere, murdered their friend, and dumped his body, after tying him up and potentially weighing him down, then back to their homes, without some signs of struggle, trauma or dirt and blood on their clothes and bodies, and got to bed by the time their parents said they did?

These were the people who'd last seen Brian alive, and so they all needed to be re-interviewed.

There were no sightings reported after eleven forty-five p.m. when a witness said he saw two boys staggering, one matching the description of Brian, in the centre of town. If it was Brian, then who was the other male? She read the descriptions. One indeed matched Brian, with his brown leather jacket and white T-shirt. The other wore blue jeans, a large black parka coat – despite it being summer – and black baseball boots. He had his hood up and was holding the man matching Brian's description upright, who seemed the more inebriated of the two. They were singing.

Brian being the gregarious soul he was, and being drunk, might well have had a sing-along with a stranger. Was he vulnerable and lost because he was intoxicated, and did somebody take advantage of him, pretending to be his buddy, who then led him somewhere to be robbed, or violated? It was a theory.

Kate took a note of the person who reported the last known sighting and dialled their number.

Mr Dougal had been a taxi driver at the time and, yes, he remembered the night clearly. Kate sighed with gratitude. She asked if she could pay him a visit for a chat. She looked at his notes, Mr Dougal would be seventy-five years old now. He told her that he was retired and she could pop in anytime for a cup of tea. He lived close to Eden House. She checked her watch and grabbed her bag, with her phone under her chin. It was walking distance.

'I offered them a ride,' he said, before she hung up.

'Did you?' she asked, stopping what she was doing. It didn't say that in the report.

'I was worried about them, I said I'd drop them home without charge.'

'And what did they say?'

'One of them was very drunk. The other said they didn't live far and they were friends.'

'Put the kettle on, will you.'

Chapter 24

Before she could get out of the office door, Dan came in and closed it behind him.

'Kate, could I have a private word?'

'Of course, but can it wait? I'm just about to pop out to see a witness.'

'It'll only take five minutes. I need your advice.'

'Oh, okay, sit down.'

She put her bag back on the floor beside Kelly's chair and waited for Dan to get whatever it was off his chest.

'Is it about work?' she asked. 'You're happy?'

'Yes, I'm chuffed to bits. The move down here did me good. I'm perfectly happy in the team, if that's what you mean.'

'Okay, so what is it?' Kate asked.

'I'm getting a divorce.'

'Right.' Kate sat stock still. 'I'm sorry, I've been through one. I know it's shit. You want me to let Kelly know in case things get rough?'

He sat down. 'That's the thing. I think the mudslinging might turn nasty. My wife has taken it badly.'

'Mudslinging?'

'There's someone else involved.'

'You're human, Dan. Work will support you, we don't listen to gossip.' Kate knew first-hand that marriage didn't suit police work. Dan was equal in rank to Kate, as detective sergeant, but not as experienced. She enjoyed being Kelly's second in command and didn't want the extra responsibility that came with being a DI. She was happy to assume the role of human resources within the team and she'd listened to plenty of personal gripes from Rob and Emma in her time.

'It's Emma.'

Kate took a breath and nodded. It wasn't exactly a secret that there was something between the two. It happened. Detectives worked stupid hours and they grew close. This looked like something more, though.

'Where do we stand, as a team?' he asked her.

Kate could see that he was uncomfortable. Dan was a dour Glaswegian who rarely showed his feelings at work. She knew why he was asking. And she also suspected why he'd chosen her for advice. She was herself currently in a relationship with the superintendent, Andrew Harris, and they'd been transparent from the word go. They were both divorced. It wasn't that the force was an ancient moral institution that banned inter-work relationships, it was more to do with performance and the potential for favourable bias.

'Ah, right. So, my advice is be honest. It's more common than you think. I'll tell Kelly, if you'd rather? Do you think it's affecting your work in any way?' she asked.

'No. Not at all,' he said.

'Just where you two decide to get a hotel?'

Dan realised that Kate was pulling his leg.

'I'm not exactly falling over with shock, Dan,' she said. 'Look, you've got nothing to worry about. In my experience, the constabulary would only become involved if your relationship compromised your work performance, and as far as I can see, that isn't a concern. Should your wife point the finger and report the both of you, the constabulary's position would be fair. You might have to answer a few questions, but we'd support you, I guarantee you that.'

'Thanks, Kate. I appreciate it.'

'Not at all. Are we done?' she asked.

He stood up. 'Sorry, yes.'

'As far as I'm concerned, that's it dealt with. You've informed us. That's all HR need, should it come to that. Gone are the days when you have to go around hiding in cupboards, which I'm assuming you don't do.' Kate smiled.

So did Dan.

'Get out of here, I've got work to do.'

He left.

She picked up her bag and left the office, walking through the incident room, noticing that Dan and Emma were working together on the initial forensic report from the cabin of Jason Cooper.

They were both superb operators, it would be a shame to lose either of them because they'd grown fond of one another. In her world, having your partner's back was priceless. There were plenty of couples in the police force who made it work – in fact, they lasted longer because each understood the demands of the job.

When she walked past Rob, she saw him concentrating on CCTV footage from the A66 around Scales, from Tuesday afternoon, when Jason Cooper drove his car home from The Swan Inn pub. He looked worn out. His wife had given birth to their second child two months ago and he was sleeping at the office more regularly.

The sound of Dan and Emma's laughter permeated the office walls, and she saw Rob peer at them longingly. She'd seen the look before, and if she were a betting woman, she reckoned that Rob was next in line to have his personal life laid bare and smashed to bits, whether he wanted to save it or not.

Chapter 25

Kelly said goodbye to Donald and Maureen. The victim liaison team had arrived punctually, and she'd left them to do their job. The couple would have had one at the time of Brian's disappearance, for a short while at least, but she guessed they'd work with the family a lot longer this time. Processing trauma was a new field of expertise within the force and they tried to provide as much support for victims as they could. It cost a fortune. But not as much as if they did nothing, in the long term.

She drove back to Eden House with a heavy heart. It was the worst element of her job but also the best motivation to get off her arse and solve the case. She'd never faced a challenge so daunting. After twenty-four years, they'd have lost witnesses, people would have forgotten details, material evidence would have been destroyed, and even Brian's body couldn't tell them what they really needed to know, though Henry was doing an incredible job. Both he and Ted agreed that the trauma to the back of the head would have caused his death. What she now had to piece together was if he was tied up before or after the killer blow, and thrown into the lake dead or alive? Brain damage could take hours to kill, even Ted had acknowledged that he couldn't give her a finite timeline. Any transfer of evidence from the killer to the victim was lost. DNA, fibres, fingernail cuttings, hair samples, fingerprints, defence wounds, sexual assault... all of it was swept away with the water. She faced a mammoth task. She'd told a white lie to Donald Miller. Sometimes extra information wasn't helpful, just cruel.

The face of Donald Miller, begging her to tell him that his son hadn't suffered, made her determined not to let him down. She couldn't help feeling that the force had done just that once before, when they couldn't find his son, and she refused to allow it to happen again.

The office was quiet.

'Kate's gone to interview a witness,' Rob told her.

'Okay. How're we getting on?' she asked.

'I've found Jason on the A66 on CCTV on his way home to Keswick at 6.05 p.m. on Tuesday,' Rob said. 'He's all over the place, boss. I'm amazed he made it home.'

Kelly stood behind him and he showed her the footage from the traffic camera. Rob had highlighted Jason's car, which was a good way in the distance from the camera angle, and, as it approached full view, it swerved sharply and corrected itself, twice.

'Jesus. I reckon he had more than two or three pints at The Swan Inn,' she said.

Kelly went to where Dan and Emma were both working.

'Morning, boss,' they said. 'Did you see Mr and Mrs Miller?' Emma asked.

'Yeah. Grim. We need everything on this. They're broken.' Dan looked stern.

'We've been working on Jason Cooper's cabin. From the photos, it was a mess, but no obvious signs of robbery, though there wasn't much to take. He was just untidy. They've lifted latent prints. Also, the Tuesday evening issue of the *Gazette* was on his table inside. It was open at the middle spread, a piece about the remains in Thirlmere.'

'I suppose anyone interested in local news would be reading about that,' Kelly said.

'But I thought we were working on the theory that he never made it into his cabin, boss,' said Dan. 'So why would he have the evening news inside?'

'Good point,' Kelly said. 'Somebody put it there.'

'They also found this,' Dan said. He brought up the collection of photographs from the inside of Jason's cabin, which had been attached to the forensic report. The one Dan showed her was of a shelving unit beside Jason's kitchen. On it were piles of mail, trinkets and general domestic mess. It was disorderly and cluttered. But she was drawn to the framed photo that Dan pointed out.

'He's in the middle of it, I think,' Dan said.

She nodded. It was an old photo, grainy, badly framed and discoloured. It looked like it had been there a long time. It was definitely Jason in the middle of the group of people. The face she stared at in science class had haunted her since yesterday morning. Five friends stood by a lake, laughing and smiling, with their arms around one another. A couple of kayaks were in the background, they'd been on the lake. To the left of Jason were Michelle and Dave. To his right, Paul and Carol.

'That was taken on our school trip after our GCSEs. It's Derwent in the background. We went kayaking.'

The five youngsters all wore wetsuits and buoyancy aids. Kelly remembered the scene from a different angle: one of a girl desperate to be noticed by the cool people.

Now it came back to her. Stronger this time. The cool group had split apart after that trip. At sixth form, they'd moved on. Michelle and Jason stuck together, but the others... dissipated. So why were they back together again by July the next year?

She remembered the fight.

'It was Brian,' she said.

'What was, boss?' Emma asked.

'It was Brian who Jason Cooper almost beat to a pulp on that school trip. It's why it ended early. But Michelle was out with him the night he disappeared and Jason wasn't.'

'We're not following,' Dan said.

'Mr Thompson,' Kelly said.

'Boss?' Dan and Emma asked in unison.

'The teacher who organised that trip. He's still teaching at Westmorland Comp.'

She walked away and went to her office. A quick phone call to the Westmorland Comprehensive, which she'd driven past this morning, confirmed that Mr Thompson, head of geography and teacher of PE, was at work today. She walked back out of her office and informed the others that she'd hold a brief at midday. She took the stairs and exited the building the back way to the car park. Uniformed officers blocked the entrance from journalists who were now two deep at the exit barrier.

'Kelly Porter!' one shouted. There was a scrum for the barrier and Kelly slammed her car door shut. She drove through the group of reporters, keeping her windows shut tight, but she could still hear the questions they shouted at her.

'Did your father bungle the original investigation into Brian Miller's disappearance?' one asked.

'Were you his girlfriend?' another threw at her.

'Is it appropriate that you're on this case?'

They'd been digging. Where did they get the nerve? Kelly felt frustrated and bemused. She drove away. The radio played a tune from the Seventies; it was oddly soothing, but then it broke for the news as she waited for the lights to change.

'It has been confirmed by the police today that the body found in Thirlmere reservoir was that of local missing lad, Brian Miller, who was last seen twenty-four years ago...'

Jesus. She slammed the digital controls and found a pop song. She gripped the wheel.

It must have been morning break when she arrived at the school, because there were kids everywhere, kicking footballs, eating sweets and wandering arm in arm aimlessly across the front driveway. There were no teachers in sight. She noticed a boy, on his own, sitting by a tree, and she wondered if he had any mates. God, she'd become sentimental. Or maybe it was Brian spending the last twenty-four years on his own in the cold dark depths of a lake that bothered her. He'd got under her skin. She parked in the staff car park and went to the reception, where she had to be buzzed in when she showed her lanyard. Security was tight. Things had changed since she was last here.

She asked for Mr Thompson.

'He's got a non-contact period at the moment.'

'Find him, please, it's fairly urgent,' Kelly said. She was aware that she was being brusque but she didn't care. The secretary made a few phone calls and tracked him down.

'Would you like to take a seat? He's on his way.'

Kelly stood.

Chapter 26

When a tall man in a suit strode into the foyer and smiled at her with an outstretched hand, she was taken aback. He hadn't changed one bit, apart from perhaps a little extra weight around his tummy. His face was warm and his eyes sparkled. It was a bit like coming home. Kelly suddenly felt sixteen again.

She'd liked Steve Thompson. He was intelligent and funny. But he also *saw* her. He gave her time and encouraged her to reach outside of her box. He was a good guy.

'Kelly! What an enormous pleasure. I always said you'd do well. Didn't you do a stint at the Met? You know your father would be so proud. You look well!'

He was just like she remembered him: assertive yet inquisitive, and reassuringly gentle. Her worries melted away. With him onside, she'd surely get to the bottom of what held the friendship group together.

'I'm good, it's strange being back inside these walls, though. Why does everything look so small?'

He laughed. She noticed now that he had deep wrinkles around his eyes, and streaks of grey hair around his temples.

'London was challenging and rewarding, but I'm glad I moved home,' she said. She didn't mention John Porter.

'Well, here you're in charge, eh? I bet it's easy to get lost in London. And you haven't got the scenery!'

'Can we talk in private?' she asked.

'Is this regarding Brian?' he asked.

Kelly looked around and nodded. 'It is.'

'You haven't changed a bit, Kelly. It's nice to see you. You went into the right profession. It suits you,' he said, and turned to lead her through a set of double doors.

She almost said 'Thank you, sir.'

He took her through another door and they entered an empty conference room. He sat down and bid her to do the same.

'Coffee?' he asked.

'Actually, that would be great,' she said. Coffee was such a distinctly adult thing, it almost jarred her to be contemplating sharing one with her old teacher.

He went to the machine in the corner of the room and fixed two. 'Sugar?'

'One. School's changed a bit,' she said, filling the gaps.

'It needed to! We have different challenges now. The internet for a start. Everything is up to date, not like it was in your day.'

He'd always been jolly. But behind the cheerful façade sat calm authority. That's why he mostly took charge of school trips; no one messed with Mr Thompson. Though that night, somebody had rocked the boat. Literally.

He sat down. 'I'm all yours. My, you've done well haven't you! Congratulations, it's always nice when an ex-student turns up and makes you proud. Not so nice when two turn up dead.'

'Two? You've heard?' She realised that keeping a lid on such events in these parts was an impossibility.

'News travels fast in Penrith. First Brian, then Jason. I've lost my share of pupils over the years. Cancer, heart attacks, prison and the like. That's life. But murder is entirely different, isn't it?'

'We haven't released the details yet,' Kelly said.

'Mr Miller, Brian's dad, has been on the TV this morning. Didn't you know?'

Kelly's heart sank. She shook her head. 'What did he say?' She'd only left them an hour ago.

'He was clearly led by the reporter, some hotshot from London. He confirmed the gossip, that Brian was killed before he was dumped in that lake, and that Jason was found underneath his cabin at Parkie's with a head injury, though the journalist didn't know that until Mr Miller told him.'

'How did Mr Miller know?' Kelly said.

'No idea, I'd presumed you told him because the cases were connected.'

'That's not the case at all. Well, that's my job. Crisis management, I'll put the press straight later. For now, I want to talk about the school trip I was on in 1996, the one after our GCSEs.'

'So you do think they're connected. People aren't stupid, Kelly. They'll put two and two together eventually.'

She felt sixteen again. Admonished, talked down to. Small. It was a tone she was unused to and one that she remembered he used liberally with other pupils.

'How did the school handle the incident between Brian and Jason? As a pupil on the trip I remember what happened, obviously, but what happened when we all got home? Was it dealt with and investigated?'

She sipped her coffee.

'It was different in those days, Kelly. You should know that. It was an in-house matter. No safeguarding, no paperwork, you know. They had a fight. Brian came off worse. It was over a girl. That was it. Jason was suspended for four days.'

'I remember Brian with golf balls for eyes, didn't anyone want to press charges?'

'They were kids. No one wants a minor to have a criminal record when it can be smoothed over in-house.' He kept referring to the old-fashioned and outdated method of not airing one's dirty linen in public, like it was an excuse for inaction. She became suddenly protective of Brian's memory.

'What happened to the girl? Michelle Parkinson came back dripping wet, screaming for help.'

Kelly recalled Michelle, crying and shaking with cold. The girls' dorm woken up, huddling around her. Then Brian covered in blood and Jason's fury.

'Drama, drama. Michelle Parkinson loved to be the centre of attention. She'd riled them up. They made up, didn't they? Or maybe they didn't, is that what you want me to say?'

Kelly noted his tone that was verging on negligent. He was, after all, talking about minors. Warning bells sounded in her head to be cautious.

'I don't *want* you to say anything, Mr Thompson.' *Sir* was on her lips as she said it.

'Call me Steve, we're not teacher and pupil now.'

'Steve.' It sounded wrong. 'I'm just trying to piece together the last year of Brian's life. You were the one to deal with them both. What did they tell you?'

He sighed. 'Brian told me he was comforting Michelle, because of something she'd told him. Jason said Brian was trying to get his girl.'

'And that was it?'

He nodded.

'Do you know what it was that Michelle supposedly told Brian?'

'No, he never told me that.'

'You taught some of these kids in sixth form, you knew them well,' she said, producing a list of the seven friends on her Toughpad. He peered at the list.

'My, they were the troublemakers. Yes, I had Dave for PE. I had Michelle and Carol in my geography class, as well as you, of course. God, I didn't miss the back of those two, I can tell you.'

Kelly smiled.

'You're not on that list? Were you not in their gang?' he asked.

'They were acquaintances. I didn't live in their world.'

'You had a lucky escape. There's a reason for that, you know. I see it all the time. The cool kids make their world look enticing, romantic, exciting, but they're often just the small town thugs and they end up in jail or dead.'

It was a harsh assessment but that didn't make it less true.

'You know your dad came to see me after the trip. He asked if you were okay,' Steve said.

It was the second time he'd mentioned John Porter.

'Didn't he want to investigate the assault?' she asked.

'We chatted about it and he agreed that it should be kept in-house.' There it was again, the safe umbrella of privacy, and knowing better. 'He said something about Michelle's father, said she was troubled, that she spread stories and caused bother. We agreed that the lads were led a merry dance. Just a bunch of sixteen-year-olds playing up. It's as old as the hills. If it can't be dealt with by school, then it spills out into the streets and that's our job, to make sure that doesn't happen.'

Kelly thought that was fair enough. After all, they were so young, and maybe she was making more of the scrap between the two boys than she needed to.

'Shocking news about Dave Crawley, weren't you two engaged?'

He knew a lot about her. 'Your dad kept me updated,' he said. 'I used to see him in the pub from time to time. I was the outside agencies liaison officer.'

'I didn't know that,' Kelly said.

'This fight between Jason and Brian really was tame compared to some of the things we dealt with,' Steve said.

Kelly nodded and finished her coffee. It was terrible, nothing like the brews she made back at Eden House. About what she expected for a school. She should have asked for tea.

'Dave got what he deserved in the end,' she said.

'Indeed. That's what I mean. Dave was the real deal, a hard criminal. Jason was a low achiever and a pain in the arse. He bummed around and failed everything. I believe he was into some petty crime, and I wasn't surprised to learn that. Well, I hope you catch whoever did this to Brian, though I suspect that somebody has already done that and taken care of it, eh?'

'We can't assume that,' she said.

'Well, in my business, dealing with kids, who are terrible liars, if it walks like a duck and talks like a duck, it usually is a duck.'

Kelly forced a smiled. It was a fair assessment of crime. Sometimes the answer was on the tip of your nose, staring at you the whole time. Complicated theories were sometimes best left to textbooks. Steve could have a point. However, his blasé assumption that this was somehow an open and shut case bothered her.

'Thanks for your cooperation,' she said.

'Oh, you don't have to be so formal with me, Kelly. I'm always here for a chat. Sorry about your dad, by the way, he was a fantastic fellow. Black and white. Old school. I don't suppose potential suspects will be hard to find, the ones on your list who aren't dead or in prison, that is. I doubt any of them made anything of themselves. Not like you.'

'It was nice seeing you again, and the school. A trip down memory lane,' she said.

'I'll miss it,' he said.

'Are you going somewhere?' she asked, surprised.

'I'm retiring after the summer, this is my last term.'

'The school will miss you. How long have you worked here?'

'All of my career. Thirty-five years.'

'Well, I'll be in touch.'

He walked her back out to the foyer, the way they'd come, and they said goodbye.

Kelly looked back; he was still watching her as she approached her car. As she turned away from him, what had at first felt like nostalgia turned to relief to be leaving the place that held so many bad memories. And now they were worse.

Chapter 27

Kate sat down and accepted a glass of cold lemonade from the old man who kept his house tidy and clean. She liked him. He had a twinkle in his eye. He told terrible jokes and chuckled to himself.

'It's just me now,' Mr Dougal said. 'And the cats. I was young then, mind you. My mind was sharp, which is why I told the police what I did, because I had absolutely no doubt I'd seen that poor boy.'

Kate had pretty much been told his life story. He'd moved here from Ireland, away from the Troubles, only to face a barrage of racism in small town England. His formal education had meant nothing over here in the 80s.

'I was lucky to get the job. They told me not to speak to the customers. I asked if I could tell them jokes.' He chuckled and so did Kate. 'It didn't go down well.'

His accent remained true to his birthplace, but Kate had no problems understanding him, having grown up around people from Cumberland all her life. She sipped her cup of tea. It was like her father used to make: you could stand a spoon up in it.

'I know it's a long time ago, sir, but we're trying to piece together the last moments of Brian's life.'

'I read he was called Brian Miller. Are his parents still alive? That's a lot of years without their son. It'd kill me. Mine are grown up now. I can't imagine them being taken too young.'

'Yes. They're local. They were informed this morning.'

'Ah, my heart goes out to them. It's down to you to make amends after all these years.'

'In what way do you mean make amends?' Kate asked.

'Well, it was a bit of a cock-up, wasn't it?'

'Can you elaborate?' Kate asked.

'Ah, now, that tells me you have no idea what went on. That's all right, you were probably a wee toddler at the time,' he said. His eyes twinkled.

Kate was tempted to be flattered, but took the statement with the affection that was intended. She was clearly older than that.

'I guess there was a lot of prejudice at the time, even in the Nineties. Oh yes, it was a white Englishman's world. I should know. The coppers weren't really interested in my statement, I knew that at the time. But I know what I saw, miss.'

'Can you remember what you reported?'

'As clear as day. It was a light night, being summer and all. The weather was beautiful. It was the only thing that kept me in England. The sunsets up here. My legs don't work like they used to but I often used to walk up Latrigg to watch it. It's like nothing on God's earth. That golden orange ball descending from the heavens.'

Kate felt like she could listen to this man all day. She waited.

'It was late. The sun was dipping down over Derwent. I was in Penrith, but I know my sunsets. I never minded working nights. It gave me time to think, and with three girls at home, it was my space.'

'I hear you, I've got three girls too,' Kate said.

'Blimey, we're in the same club. Run you ragged?'

'They certainly do,' she said.

'So, I pulled up behind these two boys because one of them was paralytic. The one in the leather jacket. He was being propped up by the other one, the one in the black parka, with his hood pulled up. Shifty sort. I knew he was up to something because he kept his hood over his face. I offered them a free ride. The one in the parka swore at me and called me a kiddie fiddler.'

Kate raised her eyebrows.

'I saw the lad's picture in the paper every day for months, until it all died down and the police gave up. It was him. The one in the leather jacket was Brian.'

'What did you do after you spoke to them?'

'I followed them.'

'Did that go in the police report?'

'I have no idea, because they never even asked for a proper statement or a signature. But I can tell by your face, young miss, that it didn't, eh?'

'Where did they go?'

'They got into a car.'

'Can you describe it?'

'Of course. I take note of these things, you see, because I kept an eye on the competition. It wasn't a taxi. It was a Ford Fiesta, an old battered red thing. Looked ancient. I know what you're going to ask me next. I got the number plate. I memorised it. I gave it to the coppers. You know sometimes in life, when you know bad things have happened but no one listens to you, you get this gut feeling that certain things are important. I wrote it down.'

Kate smiled at him. 'Nice work. Do you still have it?'

'Of course I do! I wrote it on the evening newspaper. I got it out when you said you were coming. I keep all sorts. My daughters are constantly on at me but I tell them you never know when something might be important, isn't that right? And look. Here we are.'

He got up and Kate watched him walk to the small table under the window, with a single chair next to it, maybe for watching the world go by, or perhaps for doing the crossword. He picked up an old paper, which was folded once. It was light brown in colour and she could smell the age of it as he approached. He handed it to her.

'Page four,' he said.

Kate unfolded the paper and turned to page four. She stopped at the article about the missing boy. At the top, above a photo of Brian, was a string of numbers and letters resembling a vehicle registration. Mr Dougal confirmed that it was. It was a W registration and Kate did a quick calculation in her head, coming up with 1980 as the date of production. The police checked past owners of vehicles all the time and Kate reckoned it wouldn't take long to trace one from 1980. Even if the car was now buried under a yard of scrap, and had been for twenty years, it would still have left a manufacturer's footprint, and somebody would have bought and sold it. It was a lead. She wanted to hug Mr Dougal.

'I'm assuming that you gave a good description of the parka-wearing boy to the police too?' Kate said.

'Of course. He was taller than the other lad, and well built. Local accent. I could tell by the way he told me F-off, sorry.'

'No worries, I've heard it all before, and you're just quoting,' she said. 'Ethnicity?'

'White, like Brian. And he had bright ginger or dark blonde hair, I remember thinking it matched the sunset. It peaked from underneath his hood and I thought, ah you're not all that hard, you've got sunshine plopped on your head.' He chuckled.

'Same sort of age?'

'Yes, they seemed like mates.'

'Anything else?'

'Yes, there was a girl in the back of the Fiesta.'

'A girl?'

Mr Dougal nodded. 'She looked like she was waiting for them, and a man was there too, definitely older than the kids. That's when I drove off, because I thought they'd called an older brother or something, to help them.'

Chapter 28

'Hey, Kate,' Kelly said, as her second in command came into her office. 'You look deep in thought,' she added.

Kate plonked herself down in the swivel chair opposite Kelly's desk. She relayed her conversation with Mr Dougal.

'None of it was in the original report,' Kate said.

'I'm with you,' Kelly said. 'Why would they omit such convincing testimony and not check it out?'

Kate shrugged. 'Want me to get the vehicle check done?'

'Yes, get on it now.'

'How did your trip down memory lane go at school?'

'Weird.'

'Weird?'

'I felt sixteen again.'

'Surely that's a gift, people would pay good money for that.' Kelly chuckled. 'By the time I left, I felt reassured enough that the gaggle of friends around Brian was just that. I keep forgetting they were just kids. But he did intimate that it wasn't too far-fetched to suggest that whoever killed Brian, after the body resurfaced so publicly, bumped off Jason as well.'

'It looks like it was a pretty half-arsed investigation. Who was in charge? I know your old man signed the report off, but he was a uniform following orders, who was the SIO?'

'Close the door.'

Kate did so and sat back down.

'There wasn't a detective in charge of the case, it was left to the beat on duty. That's why my dad signed the report. It was his investigation.'

Kelly had been asking herself the same questions when Kate came in.

'Jesus. So it wasn't even led by someone qualified? Why?'

'I remember my dad telling me that I was wasting my time going the detective route. He said a good copper could smell crime, they didn't need fancy qualifications to solve it.'

'I don't know what to say. That's negligence. What can we do?'

'It's abuse of process too. I feel sick. I know the force was different back in the Nineties. Cock-swingers thinking they could pick and choose who to arrest and how to keep the local hoods in check. It's one of the reasons I went to London. But it was just the same there, and still is. I don't want you to think that because of my relationship to John Porter, I won't do everything I can to get to the bottom of this. I'm beginning to wonder if he might have been given SIO status on other cases, and handled them just as badly.'

'The team will pick it up soon, they're bound to, look how quickly it became apparent to us,' Kate said.

'I know. I've asked everyone to get together at one p.m. I'll go over it then. Just the crime unit. I'll give a general brief later. I don't want rumours flying and it getting back to somebody who was involved at the time. I'm well aware that half the coppers in this county are sons and daughters of coppers from a different era. It's not their fault. But I want this dealt with properly. I need to get all my facts straight before I go to Andrew Harris.'

Kelly had no problem discussing delicate matters that involved the superintendent with Kate. She didn't tittle tattle.

'On an entirely different matter, I've been given a slot to interview Dave Crawley. I know, I know, don't look at me like that. I've decided to go myself. This case has to be done properly, and I want to see his face when he tells me about his old pals Brian and Jason. I know he couldn't possibly have killed Jason, but he could easily have had something to do with Brian's death. I completely understand if you don't want to come with me. He's been moved to HMP Highton.'

Kelly watched Kate blow out her cheeks.

Her last experience inside the walls of HMP Highton had been terrifying. A riot had called for the constabulary to place a cordon around the prison and Kate had been trapped inside.

'I'm coming,' Kate said.

As they walked through the incident room, more desks had been taken up with uniformed officers drafted in for the two murder cases.

'We've tracked down Carol Fisher, boss,' Rob told Kelly. 'Same for Tracey Dalton. They live close to one another here in Penrith. I've arranged for them to be visited at the same time, though if they've got anything to hide, then they're probably on the blower to one another right now, comparing their stories.'

'Nice work, Rob,' Kelly said. 'What about Paul Gordon?'

'That's where Dan and Emma have gone.'

'Great. We're going over to Highton. We'll be back for the brief by one o'clock,' Kelly said.

Rob threw a look at Kate and she put her hand on his shoulder. 'I'll be all right,' she said.

'I should go,' Rob said.

Kelly looked at Kate. 'It makes sense that you follow up on the forensics from Jason Cooper's place. You were there at the scene. You could have another chat with Michelle and ask her if she knew who was driving the Ford Fiesta that night too,' Kelly suggested. She didn't want to tell Kate what she should do, but equally she agreed with Rob, that perhaps he should accompany her to Highton prison instead.

'And you could prepare the brief,' Kelly added. 'I'll push it back to two p.m.'

'All right, you two win. I'll stay here.'

'Rob, get your stuff. Let's go,' Kelly said. She looked at Kate, whose shoulders seemed to release a weight. Maybe she wasn't ready to face that particular demon yet after all.

Chapter 29

Highton prison sat nestled in the middle of ancient moorland, close to Seascale, on the Cumbrian coastline, which the Victorians had deemed a perfect settlement for consigning prisoners to memory and keeping them away from polite society. On a summer's day, the drive around the beautiful coastline was a welcome diversion. If it hadn't been for the true nature of their tour, Kelly might have enjoyed it, but her stomach twisted in knots at going back there, not least to face Dave Crawley.

Dave had been her first boyfriend. Wendy and John Porter had been delighted at the match. Dave was a hard-working local lad, with a decent respectable family name and, more importantly, he knew how to charm. Kelly was swept away with the notion that somebody fancied her and, not only that, he was one of the most popular boys in sixth form. He was sporty, funny, brave and ambitious. At least Kelly had thought he was. The comments from John Porter about treating a woman right and providing for her hit a different note now that she was a fully grown woman who'd seen her own potential. John Porter's reputation at the constabulary was formidable. He'd been seen as a no-nonsense copper who caught thugs and taught them a lesson. When she'd first moved back to Cumbria four years ago, she'd been inundated with older coppers stopping her in corridors to offer their sympathy for his passing, at the same time telling her a story or two about how there was no messing about with John. She knew now that she'd need to pull some of those old colleagues in to answer some awkward questions.

She knew that people like Dave never changed. That's why she believed in the penal system for serious criminals. Many of them were unable to rehabilitate because they were wired wrong. Her concern, though, was how far had Dave pushed his group of friends to the wrong side of the law, even when they were still teenagers?

'It's always easy to see the bad 'uns at school, because the girls with short skirts and too much make-up hang around them like bad smells,' John Porter used to say. So why hadn't he sniffed Dave's badness?

She concentrated on the road.

Rob seemed distracted. She'd filled him in on what she'd learnt this morning, mentioning her concern over how Brian's disappearance had been handled at the time.

'I noticed that some basic interview stuff was missing when I looked too,' he said.

They'd all been given a copy. She knew that her team were speed readers and could trust they'd all come up with their own theories on what went right and what went wrong, it's why they worked so well together. She could tell that he wanted to give her his opinion of it.

'Like what?' she asked.

'This bloke in here,' he said, nodding at the gates of Highton.

It looked gloomy and formidable, even in the sun. Kelly recalled the throng of police vans and cordons here during the riot. The control tent. The noise when the prisoners broke out onto the roof. The smell when the police eventually got inside. The governor had been replaced, and the contract stripped by the government. It was a model overhaul, at massive public expense, but it was necessary. When the lunatics take over the asylum, something has to give.

They were buzzed in through the outer gate and parked as close to the inner gate as they could. They walked to the huge metal gates and showed their lanyards. They were buzzed through a further three sets of doors and met by a screw who seemed friendly enough but Kelly knew better. Prisons belonged to the governor, and the police were rarely welcome. Technically, Kelly didn't have the jurisdiction to interview convicts. She had to get permission from the governor. It had been granted this morning. It was like landing on an island with its own law and order.

Of course, Dave Crawley didn't have to talk to them, but Kelly, knowing the narcissist that he was, knew that he'd be chomping at the bit to have a good old chinwag with his nemesis, which is why she wanted to come in person. Dave's temper was incendiary, which was how she'd ended up getting strangled by him in an interview suite. She'd riled him.

The fact that they once shared a bed was repulsive to her.

They were taken into the visiting wing. Nobody except the screws were allowed inside the accommodation wings. They were led into a large bare room, with lots of chairs and tables, and Kelly scanned it with a trained eye; she knew that if Dave Crawley wanted to, he could easily hurt her. Kelly could handle herself, but she was glad she'd brought Rob, who, at six foot three, could protect her if he needed to, and overpower Dave. He'd be cuffed for the interview.

They sat down and the officer left. He returned a short while later with Dave dressed in grey tracksuit bottoms and top, shuffling in brown slippers, hands in front of him, held together with steel rings. He grinned. He'd lost weight, and he looked a hell of a lot older. He sat down. The officer stood three feet away with his legs spread, staring at them.

'Kelly, what a nice surprise. No, wait a minute, it's not a surprise, is it. You found poor Brian? And you want to know if I did it?'

They waited.

'David Crawley, this is Detective Constable Rob Shawcross.'

'Nice to meet you, mate. I'd shake hands, but you know, Kelly's scared of me.'

'All right, pal, stop shooting your mouth off,' the officer stepped in.

'They're so serious in here,' Dave said.

'David, on the night of the twentieth of July 1997, you took five friends for a drive to Friar's Crag near Derwent. Can you confirm that?'

'Yup.'

'You remember the exact night, twenty-four years ago?'

'Yup. You don't forget it when your mate goes missing.'

'So, it wasn't you then?'

'Wasn't me what? So, he didn't fall off the Thirlmere bridge and drown?' Dave asked.

Kelly had no way to tell if he was bluffing or not.

'Brian was murdered.'

He looked genuinely shocked.

'What's your recollection of that night?' Kelly asked.

'Can I get these off?' He indicated his cuffs.

'Not likely, pal. You assaulted Detective Inspector Porter last time you were interviewed by her,' Rob said.

Dave scowled. He looked up to a high window and Kelly wondered when was the last time he'd felt fresh air on his skin, or been able to face the sun and take a breath.

'What's in it for me?'

'Nothing,' Kelly said.

He bit his lip.

'I'm not playing a game, and I'm not going to pretend I have a deal for you, Dave. I don't.'

'He was shit-faced, I wish I'd dropped him off first,' Dave said.

'Was he any more shit-faced than the rest of you?'

'Yeah. He'd been having a heart-to-heart with Michelle down at Calfclose Bay.'

'Away from the rest of you?'

'Yeah, they wandered off.'

'Do you know what they talked about?'

'No.'

'Why was he allowed back in the group after what happened on the school trip the year before? I've always wondered that,' Kelly said, instantly regretting the personal touch.

Dave smiled. 'You always wanted in, didn't you, Kelly.'

'Just answer the question,' Rob said.

'It was a misunderstanding. Jason thought Brian was after his girl. It wasn't true. They were great mates, that's all. Brian was protective of Michelle.'

'Why did she need protecting?'

'Something to do with her old man being a violent bastard. We've all got our crosses to bear.'

Dave's reference to his own father's shortfalls made Kelly pause for a moment.

'But he was your dad's pal,' Dave added.

'What?' Kelly said.

Rob turned to her protectively and glared, willing her to keep her emotions in check.

Dave was still grinning. 'It's not just my dad who was bent, Kel.'

'Watch your mouth, kid,' the screw warned.

Dave was forty-one, but inside, cons were all kids.

'Explain to me why John Porter was bent, that's quite an accusation,' Kelly asked.

'Oh, it's "John" now is it?'

'Get to the point, David,' Rob said.

'Michelle's dad used her as a punchbag and your dad hushed it up, or didn't believe her, or whatever. Anyway, the point was they were part of the old boys' club around here, so when Brian's disappearance looked as though it might throw up questions about Michelle's home life, the case went cold.'

Dave raised his bound hands up into the air, like a space rocket, and made a whooshing noise with his lips.

'What do you mean the case flagged up questions about Michelle Parkinson's home life?' Rob said, and Kelly was grateful – she couldn't speak.

'Michelle's dad threatened Brian. Publicly. So the old fella was the number one prime suspect when Brian disappeared.'

'Wait a minute, publicly? Who heard it?' Rob asked.

Dave chuckled to himself. 'You don't know, do you? Because I'm guessing it's not in old John's report. He did it in front of Jason, and Paul, and me for that matter. Mr Thompson saw it. When we got back from the trip. Michelle was in a state and her old fella turned on Brian. Warned him to stay away from Michelle. Jason had kicked off at Brian because he thought he was up to no good behind his back, but he had it all wrong. Michelle had told Brian everything, because they were best mates. Jason read it wrong, so did Michelle's old fella. Everyone thinks Jason was the one who had it in for Brian, but it was Michelle's dad. And he's dead now, isn't he Kel?'

'Stop calling me that,' she said. 'What car were you driving that night?' she asked.

'Erm, God, it'd be my old Ford Capri. It was the only thing we could get six into without seatbelts.'

'Do you know who drove a red Ford Fiesta?' she asked.

He thought for a moment.

'Yeah, that was Jason's car.'

Chapter 30

The team assembled for the brief.

Kelly felt like she'd worked a twelve-hour shift and she was only getting started.

She felt their eyes on her, expectant. Kate had prepared a factual presentation in her absence, and after what she'd heard from Dave Crawley, Kelly was thankful to have a few Power-Point slides to bounce off.

She started with Brian, and a large photo of him appeared behind her on the whiteboard. Kate flicked the lights off and they all took a moment to absorb the image.

'Brian Miller, seventeen years old. You can see he's wearing a leather jacket and we believe that it's from this jacket that his restraints were cut. Preliminary findings conclude that he probably died of massive blunt force trauma to the back of the head. One hard blow would have done it and the wound is a single site, which would support that. It would have killed him eventually, and I've been assured that he would have lost consciousness, due to brain swelling, quickly. There were two broken ribs but it's impossible to assess whether these were inflicted before the head trauma. It's plausible that they could have been disabling blows, to get him still enough to tie up. But we do believe that he was intoxicated and so might have been easy to overpower. The rib injuries were definitely inflicted before death and not after, say by rocks at the bottom of the lake, because the break sites are the same colour as the skull fractures. If they happened later, they'd be different, that's what the experts told me.

'At some point, he was dumped in the reservoir. Now, we're working on the date of death as the twentieth of July 1997 because of what he was wearing, corroborated by his father. The last sighting of Brian was given by his friends; however, new evidence suggests that a taxi driver saw him being helped into a red Ford Fiesta by another youngster with ginger or blonde hair, number plate DPC 967W, later that night, in Penrith, and the same witness says he saw a girl in the back of the car. The car was registered to Jason Cooper. This was a tip-off from Dave Crawley and it's been verified. Jason had bright blonde hair. And that's our first problem: it looks like he's our prime suspect, but, as we all know, Jason was killed on Tuesday evening. My other problem is motive. The profiling of a teenage killer isn't usually that sophisticated, in that a seventeen-year-old may very well have delivered the blow, but to drive the body to a remote lake and dump it is highly unusual; one would expect a young man to panic and make mistakes. This has the hallmarks of extensive premeditation, and in my experience, seventeen-year-old boys don't generally operate like that. Moving a body on his own would have been difficult too, so we're perhaps looking at an accomplice.'

Kelly pressed a few keys on her laptop and several photos appeared around Brian; they were of seven of her school peers.

'These were the people with him that night, and they're the contemporary photos gained from the school, from the 1996 yearbook. I've got no doubt that there was angst within the group because of an incident on a school trip in their final GCSE year. I spoke to the teacher in charge of the trip, who still works at the Westmorland Comp, and he corroborated the narrative. Jason Cooper attacked Brian, ostensibly because he was jealous that Brian was close to his girlfriend, Michelle Parkinson.'

Kelly pointed out the kids she referred to.

'Killing him with a massive blow to the head, with no other evidence of other wounds, like in a fight or something, seems like overkill to me, boss,' Dan said.

'I agree,' Kelly said. 'Jealous boyfriend stuff doesn't cut it for me either, which is why we need to get the stories of these characters from that night ironed out. Dan and Emma, you managed to catch up with Paul Gordon, what did he say?'

'He was evasive about the whole thing, but he did say that he walked the girls home on the twentieth of July 1997, and that was the last time he saw Brian. He said he didn't see Jason that night. He did tell us, though, that was the beginning of Jason's erratic behaviour, as in he stopped coming to school and his drinking and fighting got out of hand shortly after,' Dan said.

'I remember. Jason dropped out of A Levels and Michelle not too long after that,' Kelly said.

Kelly pulled up the image of the framed photograph found in Jason's cabin.

'This was a tight group, we need to get to the bottom of why they split up. I'm going to pay Michelle Parkinson another visit. What about Carol and Tracey?'

'I've got bodycam from the visits, boss,' Rob said. Kelly cleared the whiteboard so he could play the footage for them. They watched the interview with Tracey Dalton first.

Kelly's stomach flipped over as she came on the screen. In the footage, she opened the door to the officers and smiled. Kelly studied her face and her body. Twenty-four years could be a lifetime, but it could also be the blink of an eye. Tracey looked the same in the sense that her face was still hers, but she moved and spoke like a middle-aged woman, and Kelly was shocked at the change in her. Gone was the twinkle in her eye and the sway of her hips, no longer able to lure in half of the sixth form. Her hair was fully grey. They listened as the officers explained what they were there for and they moved to a sitting room. They built up to the questions about the details. It was a ploy to make the interviewee comfortable. Tracey didn't look comfortable. When asked about the night of the twentieth of July 1997, she fiddled with her fingers and became teary eyed.

She swore she knew nothing about Brian's death. She looked like a woman in shock, having believed that he'd taken himself off over two decades ago.

'No one ever thought this is what had happened,' she said meekly.

'Can you remember anything out of the ordinary about the night?' the officer asked.

'No, I was pretty wasted. Paul took me home. I was in so much trouble the next day. I felt like crap. We never saw Brian again.'

'What was your relationship to Jason Cooper?' the officer asked.

'He was a friend.'

'Did you remain friends after he dropped out of school?'

'Kind of, for a bit. He changed. He was wild. Dangerous. Violent. We warned Michelle, but she said he'd never hurt her.'

'What did you warn her about?'

'That he was different. He was a bad 'un. She was worth more than that.'

'Why did you think that when you'd been mates for so long?'

'People change. I saw him in the pub one night, a few years after that. He was drinking with an old teacher of ours and he was just different. He looked at me but there was nothing there. I don't know if he even recognised me.'

The mention of a teacher caught Kelly's attention. The footage was stopped and Rob played the next piece. The interview of Carol Fisher opened much the same way. It could have been the same house, and the same woman for that matter. She'd aged prematurely and none of the power Kelly remembered her exuding at school was visible or even suggested. She looked as though life hadn't been kind. Kelly felt sorry for the two women. Where had their spark gone? She remembered the trio – including Michelle – as fighters, winners, triumphant Amazons of independence, like Thelma and Louise. She'd seen it still in Michelle's eyes, and in her passion for her achievements and the way she spoke, but not these two. They were beaten former shadows and it grated on Kelly's nerves. How could that happen in twenty years?

Carol gave the same story as Tracey: she went home drunk and got told off. She avoided Jason after he dropped out of school. She assumed Brian had taken himself off somewhere, and she missed him. Things were never the same again. There was emotion in both of the women's witness statements and they were believable. The footage stopped.

'Their alibis rule them both out after about ten thirty at night. That leaves Brian and Michelle alone after about eleven p.m. There's no one to corroborate her alibi – her parents are deceased – and we can't speak to Brian or Jason. Who was the teacher that Carol saw Jason drinking with? I know it was pretty common in the Nineties for students and staff to mingle but I got the impression that the teachers couldn't wait to see the back of him, so who would he be drinking with from the school?'

'That wasn't followed up,' Rob said.

'Find out,' Kelly said. 'A phone call to Tracey will do. Both women will have to come in and give formal interviews under oath, this was just to see how the land lies,' she said. 'I suspect that Carol is being evasive because Michelle Parkinson told me that Carol and Jason had a relationship after school. Let's move on to Jason Cooper.'

The whiteboard was now split in two. Brian's details and everything they'd learnt about his disappearance was moved to one side and Kelly brought up a photo of Jason Cooper, which was placed in the middle. She filled in the details: what they'd learnt from Parkie's, the witness statements from the Graham family, the fight at Connor's farm, Jason's possible dabbling in the selling of stolen farm equipment on the side, and the time he arrived home. Then she concentrated on the crime scene. She used a series of diagrams to project possible scenarios leading up to Jason being attacked from behind, then dragged under his cabin and left to die.

'Preliminary autopsy findings suggest a blunt trauma to the back of the head and a single blow with a sharp edge to the back of his neck, either of which would have killed him. He bled out. My concern is, who could drag a dead weight the size of Jason under that cabin?' She brought up a photo of the space so they could all appreciate what a task this would have been.

'Why bother, boss?' Emma asked.

'Exactly,' Kelly said. 'One suggestion is shame. The killer hid his work, which suggests to me that it was a necessary killing, not one of passion. It's clinical. Job done. Tidy up. Leave. Emotionless.'

'Are we treating these two killings as connected then, boss?' Emma asked.

'Not yet. There are similarities obviously, but the time lapse is baffling, unless the news on Tuesday morning of a body in Thirlmere sparked something and Jason needed dispatching because of what he knew.'

'That would point to Jason not having killed Brian,' Emma added.

'Bingo,' Kelly said.

Chapter 31

In the mortuary of the Penrith and Lakes hospital, two old men concentrated on their examinations. One could be forgiven for thinking they were each working on a complicated jigsaw puzzle. Their methods were the same. It was only when up close that it became clear that one peered over a partial skeleton and the other a corpse.

Henry looked up as a technician interrupted him to inform him that a selection of heavy tools had arrived for him. They were brought into the sterile room and Ted stopped what he was doing. He was looking forward to the delivery too. They each had requested the equipment to measure up against the wound sites on both victims. In some ways, Henry, with only a skeleton left, had the cleaner job, and therefore easier, but Ted had the fresh cadaver, kept in the fridge for his second night, and so more chance of matching a weapon to a wound, simply because he could see the damage to the soft tissue. They each had their expertise and the ability to aid one another in their analysis. Ted couldn't help thinking how helpful this scenario would be in training: teaching students how to approach a wound inflicted on living matter, and then examining what it would do to the skull underneath. They did it from textbooks, but this was something else.

Nowadays, such work was fulfilled by outsourced labs specialising in computer-aided design technology. The company that Henry used had sent him a software package and now all they had to do was upload images from a 3D camera, which would then reconstruct the wounds, and compare them with the edges and flat surfaces of a range of weapons. There was precedence to this kind of process. It had been used in several cases of death by gunshot wounds and injuries had been matched to items so specific as to rule out reasonable doubt. Henry had given him the example of a young man he'd worked on who'd been beaten with the muzzle of a toy gun; the software had come up with a perfect profile of that weapon. The police, in this case in Manchester, had been looking for a more sinister tool — a rifle perhaps — but the science had proved otherwise and a child's toy, upon gaining a warrant for a search, had yielded a match using all the spatial locations on the surfaces of the plaything. Photogrammetry was now an accepted form of evidence-based forensics and Ted wondered at how many leaps forward in science he'd witnessed in his forty-odd years as a pathologist.

They told the technicians where to arrange the tools and went back to their task.

Ted watched Henry use the 3D camera, which was innocuous enough. It was like a small computer mouse, which his colleague held against every surface of the gaping hole in Brian's skull, inside and out. There were probe attachments of different sizes, including one with a head the size of a small pencil, which could enter the sites of punctures in soft and hard tissues. The camera at its tip had a 360-degree range of exposure and Henry held it with one hand and controlled it with his other, using the computer. It was genius.

Ted watched on the screen as a 3D image pieced itself together into a recognisable skull.

Then it was Ted's turn.

Jason's surfaces weren't that clean, and Henry wished him luck.

'That's why I'm a bone man,' Henry said.

Ted smiled as he tried to concentrate. He'd used a machine like it a couple of times before but the sophistication of the equipment excited him. He soon got the hang of it and watched as the computer put together an image of Jason's head. He probed the lacerations of flesh and managed to get it inside the wound itself. Jason had been sent for a cranial scan, and this had already been inputted into the programme to complete the picture. Unlike the image of Brian's skull, the photograph generated by Ted's work on Jason was more detailed, as it included the damage to soft tissue around the wound site. It was like a cartoon animation and Ted understood it was the same technology used by the film industry to create realistic trauma in war scenes or – closer to home – murders in horror films.

'That's incredible,' Henry said, as he watched the image of Jason's head come together.

'Isn't it?' Ted agreed. He reached for his saw and prepared to remove the top part of Jason's head, so he could get to the brain to remove it, weigh it, and slice it into bits to send to the lab to assess the level of damage and determine decisively how long it would have taken Jason to die as a result of the brain damage. It was a standard part of the autopsy but Ted had waited until after the probe had taken its 3D pictures and now he could get inside. The scan had shown the extent of the lesions on the brain and now they had a picture of the injury site from the inside out as well as the other way around. Scanning the inside of the skull, after the removal of the brain, would be the final piece of the puzzle and show conclusively how Jason had died.

He planned to do the same on the cervical vertebrae, so he could determine which blow had killed him.

'I'm not squeamish in anyway, my friend, but this work is unnecessarily messy. Four days in my bug box and my little friends would have this cleared up, no bother, and then we could see with our own eyes what the blow did to his skull.'

Ted knew that forensic anthropologists were a beetle's best friend. Any scraps of flesh adhering to bone they wanted to examine were often cleaned by feeding them to the flesh-eating critters, like a little boy in his bedroom feeding his lizard live locusts bought from the pet shop.

The scan had proved that there was one impact site, as Ted suspected. So Jason had been hit once. They both agreed that this meant that the weapon would have to have a wide surface area. The thing with blunt force trauma, as opposed to sharp force trauma, was that it caused damage with force over penetration; the opposite was true when looking at penetrative wounds from things like knives, screwdrivers and even hammers. They were looking for something with a wide surface area, rather than a sharp end, which is why they'd requested items that could become potential weapons such as a spade, a farm shovel, a digging fork, a baseball bat, a heavy saucepan, a hoe, and even a solid metal bucket, to compare to the wounds.

'Are you thinking what I am?' Henry asked Ted when he finished sawing and began pulling Jason's skull apart.

Ted nodded.

The images of the trauma wounds to the heads, twenty-four years apart, were almost identical.

Chapter 32

Kelly drove along the A66 and thought about the people she'd gone to school with. She saw them increasingly as players in a drama, behind which secrets, currently a mystery to her, still had to be revealed.

Michelle's holiday park was a relatively recent addition to the growing geography of the Keswick holiday let industry. Cabins, caravans and campsites out of town were one way in which the industry was slowly expanding, but the beauty of the Lake District was that there was a finite area of real estate. It kept numbers manageable. Just.

The gates were open, but the police were still present, and so there were few people around. The ones that were there milled around pointing fingers and no doubt making wild accusations. Kelly parked close to the office and got out of her car. An active crime scene was kept sealed until every shred of evidence had been collected. The forensic vans had left, as had the photographers. It couldn't be good for business and she felt sympathy for Michelle, who'd made a good life for herself.

Kelly found Michelle alone in her office, looking at her computer screen.

She looked up with a scowl which soon turned to a smile when she recognised Kelly.

'Kelly! Hello. I thought it was the press or some dark tourist asking to be shown where the body is,' she said.

'Nice. You okay? It must be tough on business.'

'You'd be surprised. Actually, I've had a flurry of bookings. Sick isn't it? I thought about closing, but I need the money. Can I get you a drink?' she asked.

Kelly nodded. 'That'd be great, a coffee please if you have it.'

'I have a brand new machine, and it's just for me, though Jason did like the odd cup.'

'That's why I'm here, I thought it'd be a good opportunity to have a proper chat about him.'

Michelle nodded and went into a small room behind the desk, and Kelly heard jars being opened and water being poured. She poked her head around the door and peered in. 'This is nice,' she said. It was decked out like a small kitchen. It was clean and tidy. She allowed Michelle to make the coffees without disturbing her. The woman was deep in thought and since yesterday, her shoulders had most definitely dipped. Kelly had left her feeling that Michelle finally had her shit together, despite what Mr Thompson had said about the group of children in his care. But now, twenty-four hours later, she looked troubled. After making the coffee, Michelle dropped the latch on the office door.

'So we're not disturbed,' she said.

They sat down.

'I went to see Mr Thompson this morning,' Kelly said, diving straight in, deciding that she might as well be honest with Michelle. Kelly knew that she wasn't a woman to mess around with and would see straight through any police ploy to toy with her.

'Ugh, God, is he still a twat?' Michelle said, almost spitting her coffee.

'Why do you say that?' Kelly asked.

'Sorry, I forgot you were teacher's pet. For the rest of us he was a total arsehole. He strutted around that place as if he owned it. But you were saying?'

'I went to ask him about the school trip, in 1996, after our GCSEs. I don't recall him being a twat, but then it seems like I missed a lot,' Kelly said.

It would appear that her school memories were entirely subjective, as was only to be expected, but she felt a kind of shame sat with Michelle, who'd clearly had a different experience. Was it because she wasn't the daughter of a copper? Did teachers like Steve Thompson label kids before they even got a chance? Kelly had always thought of Michelle as an enemy, not because she was offensive or threatening towards her, but because she was in a different league at school: the untouchables. She was torn between acting like the detective, and the peer she was twenty-odd years ago. Neither felt natural.

'Can you go back to that night on the school trip when Jason hit Brian?'

Michelle closed her eyes. 'Do I have to?'

'I saw you, and I know you saw me. You were wet through, crying and traumatised. What happened?'

'Is this about Jason or Brian?' Michelle asked.

'I don't know yet. Just tell me your side of the story.'

Michelle sipped her coffee and rested her head back in the chair. There was a floor fan in the office, and it whirred around creating a pleasant breeze. It blew over them and then passed on its revolution towards the door, then came back again.

'Brian was my best friend.'

Kelly waited. The coffee was damn good.

'It was so long ago.'

'I know, anything you can remember.'

Michelle glared at her. 'You asked for a story. If you'll listen, I'll tell you.'

Kelly smiled and did as she was told.

'Brian was my best friend. He was funny, free, you know, he was a good boy. You could tell his mam and dad loved him because he always smelled of clean washing, and he always told me what he'd had for tea.' She paused and smiled. Kelly saw her eyes go glassy.

'My life wasn't like that.'

Kelly was desperate to jump in and ask questions but she knew that she needed to listen to what Michelle had to say. She sensed that this was an important story for her to tell. She concentrated on Michelle's every word so that she could make notes on her phone in her car, as soon as she left.

'It was the same for you. If I look back, all I ever wanted was to have a home to go to that was warm and clean. You always looked as though you had an army behind you, Kelly Porter. You weren't afraid of anything. I was scared of everything. That's why I always had my guard up.'

Kelly was stunned at the narrative that Michelle was constructing, but she didn't say a word.

'I wore body armour as if it were a pair of jeans. I was angry all the time, I hated people getting close to me. Except Brian. We were mates at nursery. He'd pick me up if I fell, and he showed me how to do stuff.'

Michelle paused and took a deep breath and drained her coffee and put the cup down on the table.

'I trusted him. If I ever had a problem, I'd tell him and he'd find a way to make me feel better. But when I got with Jason, it all changed. There was no room for my best buddy and I shut him out. I thought I didn't need him anymore. All I wanted was the excitement of riding in Jason's car, being taken to Derwent for a joint in the middle of the night, thieving, causing trouble, and generally dropping out of life. All the while, I saw you studying and doing well, and getting prizes and looking classy, and I hated people like you.'

Kelly kept listening.

'I realise now that I wanted to be like you. I wanted Brian's parents, I wanted your brains, and I wanted a different life. But people like me don't get that chance, so I stuck with Jason. He took me to places that took away the pain and I thought that's what I needed.'

Kelly finished her coffee and placed her cup down. She waited for Michelle to resume. She was mesmerised by the woman's honesty and sincerity. This wasn't the girl she went to school with. She was witnessing the transformation of somebody who was a rebel and a troublemaker, emerging after a long battle with herself, and Kelly wanted to know how it happened.

'I pushed Brian away.'

Now real tears formed in Michelle's eyes and fat drops fell down her cheeks. She wiped them away.

'But he knew I didn't mean it. He refused to go away. In true Brian style, he hung around and watched my back. Jason thought he fancied me and that's why it came to a head that night on the school trip, but it wasn't like that. Brian was the brother I didn't have. I told him things.'

Michelle looked at her.

'My dad was a bastard, Kelly. I mean a real bastard. He beat me with the buckle end of his belt.'

Michelle registered Kelly's horror.

'Oh, he never hit me in the face, he was too clever for that, even though he was a mean old bastard, he knew never to send me to school with bruises that could be seen. No, he always hit my back, my head, under the hair, my legs and my backside.'

'I'm so sorry, I never knew.'

'No one did, except Brian. He encouraged me to speak up. I think it took about a year for me to finally pluck up the courage to tell somebody. But then I told Jason, and he talked me out of it. That's when the trouble started between those two. Jason said he'd protect me – you know, like Rambo on steroids – and said I shouldn't tell. Things were strained between them from then on. We sneaked out – the three of us – that night, at Derwent, on the trip, in our pyjamas. Mr Thompson caught us. He sent Brian and Jason off back to the dorm and he said he'd walk me back to the girls' room.'

Kelly's gut turned over. Mr Thompson hadn't mentioned this detail.

'I told Mr Thompson everything. I was drunk, we'd been smoking and drinking. My memory of it is fuzzy. But then we heard the fight. Jason was on Brian and Mr Thompson pulled him off. I fell in the lake because I ran after Brian to see if he was okay. God, his face. I didn't speak to Jason for two weeks after that. He said that I fancied Brian and he was trying to take me away from him, but do you know what I think?'

Kelly waited.

'It wasn't about me, it was about what Jason did for that prick, Thompson. Brian threatened to snitch on him, not me.'

Kelly was puzzled.

'Come on, Kelly, everybody knows. What? You didn't know?'

'No,' Kelly said, and shook her head. 'What have I missed?'

'A bit of anything and everything. Moving bags of weed, delivering porno videos, you know, the usual stuff.'

Kelly's mouth must have fallen open because Michelle laughed at her, out loud. Kelly was torn between anger that she'd been caught out, but also shame that she had no idea what Michelle was talking about, as well as frustration because she didn't know what, or who, to believe. Steve Thompson had warned her that Michelle was melodramatic. She didn't know who to trust.

'He did it for years. I have no idea if he still does. He works at Westmorland Comp still, doesn't he? You just said yourself you saw him. My God, the stench runs deep in the sewers of the institution doesn't it? A bit like the cop shop, but maybe that's changed under you? Well, maybe he's stopped, but back in the day, it was a little side earner. Everybody knew. I told Brian that even if he told someone he'd probably get nowhere, because everybody was in on it. He was everybody's pal. Jason went drinking with him. He let him drive his car.'

'Wait, Jason let Steve Thompson drive his car?'

Michelle nodded. 'When Jason was too drunk to walk, Steve drove.'

'Did he drive Jason's red Ford Fiesta when you were in sixth form?'

'Of course. There was no law against it, was there? In the Nineties, sixth formers were always out partying with their teachers. Well, except you, eh, Kelly. Sensible lass.'

'That's quite an accusation, Michelle. A teacher supplying drugs and porn to kids.'

'Oh, don't you start. I thought you lived your life under a tree but come on!'

'Wait a minute, Michelle. Are you sure you have the right person? Steve Thompson, head of geography?'

'There's only one, Kelly. Don't you believe me? Christ I got you wrong.'

'What do you mean?' Kelly asked desperately.

'You're all the same. No one believed us then and no one will believe us now. But that's what makes the world go around isn't it? Those with the power hide behind their uniforms and their proper jobs and the rest of us get given the dregs, and climbing out of the shit tub is impossible. Jason never got out of it. But you did, because you're part of the system and your dad was as bad as they come.'

Kelly sat motionless and speechless.

Michelle seemed to calm down and gather herself. 'I'm sorry,' she said.

Kelly looked at her. Scenarios whirred inside her head. What Michelle had just told her would give a plausible motive for Brian's murder, to shut him up, but that would mean her prime suspect was a person of authority, a teacher no less, and the violence that was unleashed on Brian's body didn't fit with somebody who had no criminal record, and somebody who was still teaching in the same place, still in charge of vulnerable kids. This could blow the case wide open. A churning feeling gripped her stomach.

Michelle was melodramatic...

Then there was the other scenario, and one that she knew Steve Thompson would put forward in his defence.

What if Michelle Parkinson was lying?

'That's the thing, isn't it, Kelly? A child's word against a teacher's? Your dad believed he was a decent chap. Everybody did. Just like everybody believed that my dad was.'

Her words stung. It was as if she'd read her mind. Kelly fiddled with her bag and brought out her Toughpad.

'Can I take a formal statement?'

'About what?'

'Steve – Mr Thompson. It's historical misconduct at the very least, and it needs to be investigated. It might well have pertinence to our current inquiries.'

'You always used fancy words, Kelly. Aye, I'll give you a statement if it makes that bastard uncomfortable. Things are different now aren't they? I saw in the *Daily Mail* recently that a father was given a prison sentence after hitting his daughter after a night out. Good lass. I wish it was like that back then.'

'I'm sorry it wasn't.'

'I reported him, you know. My old fella. Me and Brian went to the cop shop. It was all old men, of course. Your dad told me to start respecting my elders and sent me home.'

Kelly felt sick.

'Did you have bruises?'

'Course I did. They took a while to come out, mind. I was tough. I had a massive black mark across my arse, they didn't want to know. They said domestic fights were part of growing up if you didn't listen to your dad.'

Kelly's blood boiled. The guilt of her ancestral colleagues weighed her down. She'd seen it time and time again. But John Porter's obvious negligence made her rage with shame.

'Why haven't you gone to the police before now?' Kelly asked. 'I mean about Mr Thompson.'

'Same reason. I knew I wouldn't be believed. Jason's death brought it all back, and seeing you. I trust you, Kelly. You'll see justice done. I know it. Or at least that's what I hope.'

It was a hefty responsibility, and now wasn't the time to tell Michelle that historic cases were almost impossible to prosecute because of the lack of evidence. She changed the subject.

'The night Brian disappeared. We've had a new witness come forward. He said he saw Brian with another boy wearing a black parka, and blonde or ginger hair. It was gone eleven thirty, so after you said you'd all gone home. He also said he saw them get into a red Ford Fiesta and there was a girl in the back seat. Have you any idea who that might have been?'

'Aye. It was me.'

'What? Why didn't you tell the police at the time?'

Michelle raised her eyebrows.

'Same reason? You'd had a bad experience and didn't think your testimony would matter?' Kelly answered for her. Michelle nodded.

'So where did you go?'

'They dropped me at home because I started screaming and causing a scene. I didn't want Steve in the car.'

'Mr Thompson?'

Another nod.

'He was driving. Brian and Jason were paralytic.'

'Do you know where they went?'

'No idea, but wherever it was, Steve took them there.'

Chapter 33

Kelly marched into Eden House in a bad mood. Being played sat uncomfortably with her. Being lied to was far worse. Steve Thompson had questions to answer. She tasked a uniform with the job of arranging to bring him in. She could have easily sent a squad car to Westmorland Comp, to embarrass him, but, though tempting, that wasn't her style. Maybe that's what Michelle wanted. It seemed her old school peer had an axe to grind and was using her to wield it.

A historic misconduct accusation could ruin Steve's career, so she needed hard facts. Like his whereabouts around six p.m. on Tuesday night.

She'd ceased to be shocked at how far people would go to cover their tracks once they went down the rabbit hole of crime. It was like a game of dominos. Once you went over the edge and committed an act that could put you away, going further to cover it up was not such a big leap.

Nothing was off the table. Hadn't Steve Thompson already hinted to her that whoever killed Brian had panicked at the discovery of his body and bumped off a potential witness? Were there more? As she was once told in the early days of her detective training: as an investigating officer of murder, you knew two things. One was that you didn't do it, the other was that your partner didn't either. Everybody else was a suspect. That included Michelle Parkinson. Kelly had to put her feelings to one side as she grappled with the case. Everything she knew about Steve Thompson and Michelle Parkinson had to be shelved until she finished the job.

She took the stairs and found Kate with Rob in her office. They looked grave.

'Everything okay? I've had a breakthrough,' she said breathlessly.

Rob got up and walked past her sheepishly and closed the door.

'What's going on?' she asked Kate.

'This morning I had Dan in your office telling me he's getting a divorce.'

'Oh, and what has that got to do with Rob?' she asked.

'He's just come in and told me he's getting one too. He's been kicked out by his wife. I told him to go and pack a bag and check into a hotel but he's refusing. He says he feels better at work.'

'Bloody hell.' Kelly sat down heavily.

'What's your breakthrough?' Kate asked.

Kelly told her what Michelle Parkinson had told her, and about her meeting with Steve Thompson.

'How trustworthy is she?' Kate asked.

'I haven't a clue. I'd back him, but then that's Michelle's point, she's saying the institution closed ranks to protect themselves.'

'Why hasn't she come forward before?'

'Exactly that. She said she knew she wouldn't be believed because she was a loser who dropped out of school and he was a respected teacher, and pals with the local police department,' Kelly retorted.

'Local police department?'

'John Porter.'

'Oh.'

'Kelly, can I be candid?' Kate asked.

'Of course, that's what I rely on you for,' Kelly said.

'Is this too close to home? You're personally involved with each and every one of our persons of interest, and the victims, as well as the lead investigator. I know you can remain objective, I told you that earlier, but are you?'

From anybody else, Kelly might have let her ego take offence and throw Kate out of her office, but she realised Kate's true intentions and nodded.

'I get the distinct impression that I'm being played here. I don't know by who, or even why. Maybe it is just because I have this connection with all of them. Maybe it's because John Porter messed up. God, Kate, I always looked up to him. At least to his reputation.'

'I've got your back. I have absolute faith that you can do this without getting too close to the mettle. All you can do, if you stay on this case, is check out every lead and do what you do best. The evidence will speak for itself. Maybe not straight away.'

'I can start by checking out Michelle's story. Budge up,' Kelly said, swinging her swivel chair next to Kate's in front of her computer. She switched it on and logged on. 'Back in 1997, Michelle's report would have been filed on paper and cardex. I have no idea what retention times are in the constabulary, do you?' she asked.

Kate shook her head.

'Michelle's complaint might not have even made it to a file. If it didn't then it's dead in the water,' Kelly said.

She tapped some keys and Michelle's name popped up on the system. Kelly couldn't believe her luck.

'Jesus, there are two incidents here. I've got a 999 call from 1995, and a formal complaint made in person, here at Eden House, in 1997.'

'Whoever had the task of updating the system did it well, must have been a woman,' Kate said.

'Michelle didn't tell me about a 999 call,' Kelly said.

The call had been logged at seven minutes past midnight, on the tenth of September 1995. Kelly read the transcript aloud. The complainant was logged as Michelle Parkinson of Barrow

Road, in Penrith. The call handler's notes were typed up and on record.

> ...assaulted by her father... Mr Kenneth Parkinson... Officers visited address... Mr Parkinson drunk and uncommunicative... matter resolved... passed to social services for follow-up...

There were no follow-up notes, and this is where departments let themselves down before the digital age. Kelly picked up the phone and dialled the number for social services here in Penrith and got through to the records department for closed cases. It didn't take long and she replaced the receiver.

'No file on Michelle Parkinson.'

Next she read the transcript for the incident in 1997.

> Accompanied by Brian Miller, aged 17... assaulted by father... no evidence of bruising... history of domestic events... no follow-up... all charges withdrawn...

'Well, well,' Kelly said, sitting back in her seat. 'It corroborates that part of her story at least.'

It had never been passed to an investigating officer. She felt relief that John Porter's name didn't appear but she also experienced the deep pull of disappointment and injustice that merely twenty-four years ago, a frightened girl couldn't get the protection she needed and deserved. People thought that monsters hid behind bushes, but in reality, they lived at home.

'I'm going to apply for a warrant for Steve Thompson's address,' Kelly said.

Kate agreed. Under the Police and Criminal Evidence Act of 1984 (PACE), warrants could be granted when the police suspected they could find evidence of a criminal act being committed. In this instance, not only would she instruct officers to search for pornographic material, but also anything connecting him to schoolchildren outside the remit of his job, as well as garden tools, like a spade.

She sent the application to a local judge in Carlisle who covered their legal work. With a bit of luck, she'd get it back late this afternoon.

'Let's also go over the forensic report from Jason's cabin and the results on his car.'

Dozens of prints had been lifted from the inside. 'If Steve Thompson drove Jason's car when he was seventeen, why not now? People who keep secrets together tend to stay in touch. Your taxi driver, Mr Dougal,' Kelly said.

Kate nodded. 'Let's see if he recognises Steve Thompson as the man who might have helped the two boys get into Jason's car that night,' she said.

'We need to come up with a timeline of events ready to present to Thompson,' Kelly said. 'Can you get on that, Kate?'

Kate agreed and left the office.

Most perps, faced with the science, caved and admitted defeat. Some didn't, and it was the tricky ones who thought they could get away with it who gave them the biggest headaches. If Steve Thompson was used to keeping secrets, then he'd be prepared with a cover story, perhaps one that he'd been working on for twenty-four years. But still she found it hard to believe. Or more accurately, she didn't want to believe it. The notion that Mr Thompson, the man who'd encouraged her, praised her and given her a role model of sorts, could be side-lining in the supply of contraband to kids. It didn't add up.

Kelly opened the HOLMES app on her Toughpad and inputted the new information she'd learnt this afternoon. The dynamic reasoning software did all the cross-referencing and brain work that it took a dozen coppers months to achieve, and if it identified a match, it'd tell you within seconds.

It flashed up a notification.

It was Jason's red Ford Fiesta, recognised by the number plate given to them by Mr Dougal, who'd kept his newspaper from that day, all these years. There it was in black and white. DPC 967W.

A witness had spotted the car near Thirlmere on the night Brian disappeared, around midnight. It had never made the case file, because Thirlmere was miles away and hundreds of people rang in during the months after the disappearance of a young local lad of tender age, and most of them were dismissed as nutters. But the digital footprint remained because every call had to be logged.

'Well, well,' Kelly said, staring at the screen, stunned.

Chapter 34

'Interview with Carol Fisher. Time is four thirteen p.m. For the purposes of the tape can you confirm your name, address and date of birth for us Carol?' Kelly asked.

Next door, Kate was tasked with taking Tracey Dalton's formal statement in interview suite two. In interview suite one, Kelly faced Carol Fisher across a plastic table, audio equipment between them, a coffee cup, a carton of orange juice, and the obligatory box of tissues.

There'd been no pleasantries, like there had been with Michelle. Carol had scowled at her when she'd been brought into the room. It was as if Kelly was causing a great deal of discomfort to a lot of people, and Carol, for one, didn't appreciate it.

Kelly was unfazed. The thing about justice was that it made more people nervous than it liberated. She was used to dealing with the traumatic side of the law: the side that hurt people.

Carol confirmed her details and they got started.

'Carol, can you run through the events of the night of the twentieth of July 1997 for me.'

Carol rolled her eyes and sighed.

'Can you speak clearly into the mic please,' Kelly asked politely.

She found Carol's immature manner tiresome. She behaved like a teenager. But Carol wasn't a girl anymore. She was a woman. Of all the girls in the cool group, Kelly had disliked Carol the most. She'd been vicious. She'd enjoyed humiliating other girls in class, and had wielded her kudos of being a girl boys flocked around like a knife. It was tragic. That sort of behaviour belonged to a fifteen-year-old. Carol was forty-one.

She chewed gum loudly and wore a thick layer of make-up.

Carol ran through the night in question and gave the same standard text that she'd given at the time.

'Okay, Carol, so where was Jason Cooper?'

'No idea.'

'Were you aware that he met Brian later on?'

'No. Who told you that?'

Kelly ignored the question.

'How many lads shifted contraband for Mr Thompson? Was it just Jason? Or did you help out?'

Carol looked at her, non-plussed.

'Contraband, Carol. Illegal stuff. Black market items such as drugs, porn videos, things like that,' Kelly said.

Carol's body stiffened.

'I told him not to! I only helped him a couple of times. It was just a laugh.'

Kelly's stomach tightened. So, it appeared to be true. She just didn't want it to be.

'All right, I hear you. Did you help him deliver them or simply sort them?'

'I went with him in his car.'

'The red Ford Fiesta? This one? For the purposes of the tape, the witness is being shown a picture the same type as the car registered to Jason Cooper in 1997.'

It was a stock photo.

Carol nodded.

'The witness has identified the vehicle used to move contraband.'

'Am I in trouble?' Carol asked.

'You're helping with our inquiries in a serious offence, Ms Fisher. You're not charged with a crime.'

'Okay. He delivered all over the bloody place. I spent hours in that car helping him. He wouldn't buy me a curry until we were done.'

'You told my colleague earlier today that you had little to do with Jason Cooper after college.'

'It was a long time ago. I didn't lie, you've just reminded me.'

'How did Michelle Parkinson take to your relationship with her ex-boyfriend?'

'How do I know? They were over by then.'

'Barely, we're talking about Christmas 1997 aren't we?'

Michelle had told Kelly that Carol moved in on Jason three months after Brian disappeared and her own relationship with Jason fizzled out. Kelly had pieced together a timeline of events with the help of Dave Crawley's testimony and Michelle's statement. It was amazing how much could be drawn out of someone who thought theirs was the only story being told. It was all innocuous stuff, but that wasn't the point. She was overwhelming Carol with trivial information, so by the time they got to the heart of the matter, she was more likely to be truthful.

Kelly moved on. At the moment, she had plenty of low-level criminal activity but nothing connecting the dots.

'The school trip. 1996. Mr Thompson made sure Jason wasn't prosecuted for the assault on Brian because he knew that Brian had found out about the illegal trade he had going and was threatening to snitch.'

'Really? Wow, what an idiot.'

'Who? Brian or Mr Thompson?'

'Mr Thompson, obviously. I liked Brian. He was funny.'

'Yes, he was,' Kelly paused. 'Which is why I want to know who killed him.'

The penny dropped. 'You think Mr Thompson did it? Blimey. And I thought it was Jason.'

Carol realised her blunder. Her cheeks went pink. Even under the make-up, Kelly could see.

'Talk me through that, will you.'

'I… I… Just thought Jason had it in him. When we didn't see Brian again after that night, and I knew Jason was making it up that he couldn't come out because he still had beef with him over Michelle, I just thought it was him who did it. He was violent. He killed cats.'

'Did he?'

Carol nodded enthusiastically.

'I saw him torture them. Michelle thought it was funny but I didn't.'

'You saw this and still dated him?'

Carol shrugged.

'Did you see Mr Thompson driving Jason's car?'

'Yeah, so what? Jason was always stoned or drunk. I suppose Mr Thompson was doing the right thing in a way. Jason would have got into a lot more trouble if it hadn't been for him.'

Kelly sat back.

Witness testimony was notoriously tricky when it poured out of someone's mouth so freely. She had to compare it to what else she had. She left Carol stewing over what she remembered and turned off the tape. She took the lift up to the incident room and joined Rob at his screen, who'd been monitoring the two interviews.

'She's got a big gob, boss,' he said.

'Hasn't she just? How's Kate getting on?'

'Similar story. Tracey has admitted to knowing about the shifting of illegal gear, though she hasn't gone as far as to point the finger at Jason or Steve Thompson directly. All conjecture at the moment.'

'And they're not great witnesses,' Kelly pointed out.

'Agreed. We'll need to prove that they haven't got something to gain from framing a respected teacher.'

'Emma?' she said across the room. It had been Dan and Emma who'd visited Paul Gordon at his home to take a statement. Apparently he was between jobs. 'What did Paul Gordon say about Thompson?' Kelly asked. Emma got up and joined them. Dan wasn't at his desk, and Kelly was suddenly aware that her team had lives. And two had divorces to sort out. She remembered that Kate had been the one who both men chose to confide in.

'He didn't mention him, boss,' Emma said.

'Did you ask him about the school trip?' Kelly asked.

Emma nodded. 'He said it was between Jason and Brian, and that was it. Scraps were common within the group, he played it down.'

'We need to get this lot to realise that they have more to gain by turning on one another than they do by sticking together, and to a story that hasn't stood up to twenty-four years of being buried in the mud at the bottom of a lake,' Kelly said.

'It's a long shot boss, with no physical evidence,' Rob said.

Kelly knew that she couldn't put off talking to Paul Gordon any longer. And it was something that she had to do alone.

'I'm going to finish up with Carol, then I'll head over to Pooley Bridge to put some pressure on Paul Gordon. Rob, can I have a private word when you're finished with Kate's interview?'

'Yes, boss. It looks like she's finishing up,' he said.

Kelly watched the screen and saw that Kate was leaving interview suite number two. They waited for her to reach their level and Kate walked back into the incident room and exhaled deeply.

'She's a piece of work. She's not likely to give us anything anytime soon,' Kate said.

'Same for Carol. I'm heading off to see Paul Gordon and then I'll probably head home. Tomorrow we should have the forensic reports and I'm going to meet the underwater search team at Parkie's lake. You want to have a go at Carol and I'll pay Tracey a visit before I head off and then we can let them go?' she asked Kate, who agreed.

'I'm thinking we try throwing out what Dave Crawley told me and Rob at Highton this morning,' Kelly said.

Kate listened.

'I want to know if any of it rings true,' Kelly added. 'Let's find out if they know anything about Ken Parkinson threatening

Brian in public, and also what they remember about Michelle and Brian having a cosy chat on their own at Calfclose Bay, the night Brian was last seen,' Kelly said.

'What about the other thing, boss?' Rob said.

'What's that?' Kelly asked.

'About coppers being bent,' he said.

'Sure, let's throw that one in the mix and give them an excuse not to cooperate,' she said.

'He didn't mean it like that,' Kate said.

Kelly saw a warning look in Kate's eyes and knew it was from a friend. Which is exactly why she was heading home soon. Her head was full to bursting. But she'd already decided there was no way she was handing this case over to anyone else.

'Sorry, Rob. What I meant was, let's be subtle. Kate, let's ask them if they were interviewed by the SIO at the time, and how that went. We've got brief statements as you might expect from kids, let's see if they match their verbal recollections,' Kelly said.

'Agreed,' Kate said.

They walked to the lift together.

'Sorry about that,' Kelly said to her second in command.

'No worries, it's my job to be the terrier at your heels. Rob worships you, we all do. I don't want you losing your head on this one.'

Kelly stopped and turned to her, pushing the button for down.

'I'm under no illusion that the force was a different beast when John Porter was in charge – though, of course, as a sergeant, he shouldn't have been. If I need to drag up other cases to investigate his professionalism, then I will. What bothers me most is that he apparently went drinking with my old teacher, and everybody apart from me seemed to know about it.'

Chapter 35

Interview suite two was an identical mirror image of suite one, except there was a different ghost from Kelly's past waiting for her.

Tracey's face dropped when she saw who walked in.

'Afternoon, Tracey,' Kelly said.

'I've seen you on the telly, Kelly Porter,' Tracey said.

Kelly almost expected to be asked for her autograph. She sat down and restarted the tape, having agreed with Kate what she'd be asking Carol next door.

'You look all serious, Kelly. I always knew you'd get a good job. Not like the rest of us, eh?'

Kelly was becoming tired of the constant references to the 'girl done good' story, and the more she heard it the less it sounded sincere. The fact was, she never liked any of these people apart from Brian because he was funny, and Jason because he was handsome. It turned her stomach now to think that she'd fantasised about a boy who, by all accounts, had turned out to be violent and nasty. But then she was only seventeen.

'Tracey, I want to take you back to the twentieth of July 1997, and Michelle and Brian's private chat away from the rest of the group, at Calfclose Bay.'

The bay was a tourist magnet and a favourite of Kelly's in the autumn when the visitors disappeared and the colours changed. The view down to the jaws of Borrowdale were unparalleled and she'd sat there on the bench, dedicated to some admirer of the fells years ago, staring at the view, often at dawn. It was a place for lovers. But it also offered privacy for other more questionable pursuits. They'd caught plenty of skinny dippers down there over the years, and it was ideal for illegal transactions in contraband. Cars couldn't get down there, but it had good access from the backroads of Keswick, to the north of Walla Crag. Hundreds of paths descended off the crag down to the lake, and you could be forgiven for thinking you were in a dramatic wilderness with no one else for miles in the stillness of a Derwent night.

'They were best mates,' Tracey said, and Kelly noticed the melancholy in her voice.

'So it was an amicable moment? Just two mates going for a walk?'

'No, they were arguing. That's why it was so sad that Brian went missing after. Michelle's last words to him were nasty.'

'Do you know what they were? Those last words?'

Tracey shook her head.

'Is that why Ken Parkinson threatened Brian?'

Tracey stared at her. 'I guess so. Everyone thought that Michelle's bruises weren't from falls, it wasn't just me. Brian said he hated him and was going to the police, but he hated the police too, because they never believed Michelle.'

'Any officer in particular?'

Tracey looked embarrassed.

'You can tell me, Tracey. It's my job. I have no personal involvement with what went on in the past. Like you, I was at school, minding my own business.'

'Right. It was your dad, Kelly.'

'For the record, John Porter?'

Tracey nodded.

'Can you just run through that for me? I'm aware that Michelle and Brian approached John Porter, do you know what was said?'

'Yeah, they told him that Michelle's dad was a bastard and he beat her. Your dad wasn't interested. He told them to go home. It made Brian even worse and the injustice of it all ate him up. He wanted to take Michelle out of that situation. But no one listened. They even told the school but they said there was nothing they could do about it. They said they'd have to go to social services.'

'Who at school told them that?' Kelly asked.

'Mr Thompson. He said he was sorry but these things were almost impossible to prove. He even had a word with Michelle's dad, but because he denied it and came up with excuses, nothing was ever done.'

Kelly tried hard to hide her disappointment in the system. But she knew that Tracey could see her shame. Of the two, Tracey was the more thoughtful. She seemed to have a conscience.

'I'm sorry,' Tracey said.

'About what?' Kelly asked.

'It can't be easy knowing that if your dad investigated stuff properly, people might still be alive,' Tracey said.

That stung. Kelly stared at the tape whirring round. It was on transcript now forever. There was nothing she could do about it. The accusation would now have to be passed to internal affairs to be properly investigated and she could wash her hands of the burden. John Porter was now on his own. It was a relief of sorts.

'Did you witness Ken threaten Brian?'

'Yeah. We all did.'

'What did he say?'

'He said he was going to rip his head off and shit in his neck. We thought it was funny at the time, so did Brian. But then he disappeared and we were all terrified.'

Kelly knew the quote came from an old film. 'Did you tell the police?'

Tracey gave a look as if to ask if Kelly was absolutely off her rocker. Kelly smiled weakly.

'The witness has indicated that she was too scared to tell the police because of previous dealings with reports of domestic abuse being ignored.'

Tracey smiled and nodded.

Kelly turned off the tape.

'That was what it was like back then. No one trusted the police, especially us girls.'

After she'd told Tracey she was free to go, she met Kate back upstairs in the incident room, who told her much the same story from suite one. Finally, they had two people's statements they could corroborate. Of course, the truth was still hidden behind a myriad of issues, and the two women could have easily planned their scripts in advance. But it was more flesh for the bones.

'Did anyone manage to pin down Steve Thompson?' Kelly asked.

'He's not answering his phone, boss,' Rob said.

'Can we have that chat now?' she asked him.

He followed her to her office and she asked him to sit down and close the door.

'You okay?' she asked.

'I take it Kate has told you?'

'Yes, she has. I'm sorry, Rob. Do you need some time off?'

'No. I want to be here. Jesus, this is the only place I feel as though I contribute anything of fucking value. Sorry, boss.'

'No apology needed. You hear me swear all the time, it feels good. If you want to take your anger out on the job, I understand, but if it gets too much, then just shout. I'm going to see Paul Gordon.'

'It's catching, isn't it?' Rob said.

'What?' she asked.

'Not being able to hold down a normal life. My wife doesn't understand. I guess it was the same for Dan.'

'I think it was slightly different for Dan,' Kelly said.

Rob laughed.

'The important thing is you tell me if you're going under, the last thing I want is for you to get so stressed out that you burn out. I need you, but I also value your health.'

They walked out to the incident room together and Kelly looked up at the wall clock.

'Finish up, everybody. We'll see you bright and early in the morning. We'll have a busy day as I'm expecting a glut of lab results. Go home and take care of yourselves. Perhaps we'll get some answers from Steve Thompson by then too.'

She glanced at Rob, who looked at his watch, and she knew that he wouldn't be planning to go home tonight. Kate winked at her. She'd made a bed up at her place for him and she wasn't going to take no for an answer.

'Dan, can I have a quick word?' Kelly asked.

They went to her office and she closed the door behind him.

'Kate told you then, boss?' he said.

'Of course she did, and you could have too. We've got your back. I'm sorry. Break-ups are always messy. Are you okay?'

'In terms of work, yeah. I'm grand. It was inevitable to be honest, boss. We shouldn't have got hitched. She never wanted to move here in the first place,' he said.

'Well, I'm glad to hear you're okay, and if that changes I want to know. I've just told Rob he's to put his health first. As far as I'm concerned my door is open and I value your honesty,' she said.

'Thanks boss.'

'Right, I'm going home. My head is mush on this one. I need a massive glass of red wine.'

'Red wine in the summer, boss? That's disgusting. A pint of shandy and a packet of crisps'll sort you out.'

She slapped him playfully on the way out of her office, saying her goodbyes as she walked to the lift.

Her stomach churned at the thought of seeing Paul Gordon, but it was nothing compared to how she felt about John Porter.

Chapter 36

The last time Kelly had seen Paul Gordon was at Dave Crawley's trafficking trial. She'd been amazed to discover that he lived five minutes away from her house, in Pooley Bridge, but even more shocked that they never crossed paths. Paul hadn't been implicated in the inquiry into Dave's affairs, but had supported his pal on every day of the trial, throwing daggers to the prosecution, including Kelly on the day it was her turn to give evidence.

It wasn't that she enjoyed putting people away, it was her job. She didn't judge those who broke the law on a personal level. To her, it was black and white. There was little mitigation that could convince her either way. Dave Crawley was a perfect example. He'd been her fiancé, but she still put him behind bars. But Paul had thought that her engagement to his friend stood for something. It didn't work like that. On the stand, it had come up. Under cross examination the defence had insinuated that Kelly's past relationship with the defendant caused her to become passionate and muddled, and bungle the inquiry – in other words, she wasn't fit to be in charge of the case.

Perhaps like this one.

She drove out of Penrith and hadn't got far when Kate called her.

'It's a negative from Mr Dougal on the ID of Steve Thompson being the adult he saw next to Jason Cooper's car,' she said.

'Damn. Really?'

'He was busy looking at the drunk teen, he can't be sure. It wouldn't stand up in court at the moment, what about a line up?' Kate suggested.

'Can we arrange it for tomorrow?' Kelly asked.

They agreed that it would be a perfect opportunity. The police couldn't force persons of interest into a line up, they had to do it of their own free will.

'I'll ask him,' Kelly said. It was tempting to think that her past relationship with Mr Thompson would play in her favour, but experience with Dave and Paul told her not to trust the past.

They hung up.

She drove around the roundabout over the M6 and headed to Stainton and on to Pooley Bridge. Her phone rang on hands free. It was Johnny.

'Busy day?' he asked.

'Yep, I'm on my way home, just one more call, and it's in Pooley Bridge,' she told him.

'Chinese? Ted and Henry are here. They're like a couple of mad professors comparing notes,' he said.

She smiled. It was a welcome diversion. Despite Ted and Henry's conversation being predictably about corpses, she didn't mind. Anything was a welcome relief from her old school acquaintances.

'Sounds perfect.'

She asked him to order for her. Last time they had Chinese takeaway, Lizzie had happily chomped on noodles, prawn toast, and sucked the life out of the pork ribs.

'ETA?'

Johnny still used army jargon, partly to wind Kelly up, but also because in her line of work, she understood it.

'About seven, I reckon.'

They hung up.

Kelly stretched in her seat. She entered the small village of Pooley Bridge from the north-east and found the correct street from memory. Maybe this was why she hadn't bumped into Paul Gordon, because it was slightly away from the main drag, and Kelly, even though she lived there, didn't make a habit of wandering far from the centre. She wondered if he'd stopped drinking in The Crown because that's where she'd last seen him.

She'd had a uniformed officer call ahead, and Paul was expecting her. She parked and locked her car, walking to the front door. Her heart raced but she told herself to calm down. It was just another witness interview, and she'd done hundreds, if not thousands, of them.

Not with school mates though…

She knocked. It didn't take long for her to hear footfall behind the door. The house was a pretty terrace at the end of a row, tucked away up a hill behind the village, on the way out to Martindale. It must be hell in peak season, she thought, with the traffic going along the south coast of Ullswater. She wondered if Paul knew where she lived.

The door opened and he stood holding it, with a broad smile.

'Kel. It's been a while,' he said warmly. 'Come in.'

She relaxed a little. He hadn't changed. He earned the nickname 'Flash' Gordon because of his size and fearless courage. A fickle and immature gag at the time, but it stuck. He was tall and broad with blonde hair, and his smile was infectious. He looked as though he was still a hit with the opposite sex.

'Thanks, Paul.'

She went into his hallway and noticed that it was well kept, clean and smelled of fresh laundry. It warmed her to see him house proud.

'What are you doing these days?' she asked.

'Plastering. Keeps me busy,' he said vaguely. 'Do you want a cuppa?' he offered.

'I'd love one. You know I just live around the corner. You're my last stop of the day.'

'I did know, yeah. I've seen you a couple of times at the Chestnut Tree in the morning. It was on the news too. I saw the press camping outside of your house during the Ian Burton trial.'

'Hmm. That was interesting. I'm crap at being on TV, that's what we've got a publicity department for.'

'I don't know, Kel. You looked good to me. You should do one of those true crime things for Netflix.'

'Sod off,' she said. She followed him into a small kitchen at the back of the house and saw that his window overlooked stunning views of the fells behind Pooley Bridge.

'You've got a beautiful spot here,' she said.

'I can just walk out of my back door and go for hours.'

'Clean living?' she teased him.

'We all have to grow up sometime don't we? I've got a daughter now, I need to think of her.'

'Congratulations. I didn't know.'

'Neither did I. It came as a bit of a shock. I had a fling ten years ago and boom, out of the blue. Paternity test confirmed it, she's mine. There's a photo over there,' he said.

She turned to see where he was pointing and saw a framed photo of a small girl, laughing and blowing bubbles.

'She's cute,' Kelly said.

'Shall we take our tea outside? It's a gorgeous evening,' he said.

He opened the back door and they went out onto a small terrace. It was wonderfully private.

'How long have you been here?' she asked.

'Five years or so.'

Kelly got the impression that they'd run out of small talk. It was nice to catch up but that's not why she was here.

'I appreciate you talking to my officers this morning. I've had time to catch up on what you told them.'

'I expected you'd want to check for yourself, you're in charge,' he said.

They sat down and Kelly sipped her hot tea. It was idyllic sat here, there was no sound of traffic and the fences were high.

'You told them you didn't see Jason the night Brian went missing.'

He nodded.

'But you indicated that he started to behave oddly after that time. What did you mean?'

'His drinking got serious. I mean, we all drank in college, and smoked the odd joint.'

'Odd? I'd say it was a bit more than that,' she said.

He nodded. 'Okay. But he was aggressive. You know, off the charts. He had a death wish almost.'

'Almost?'

'He'd drive to the edge of the passes and threaten to go over, with the girls in the car, making them scream – he loved that kind of shit.'

'Nice. Carol told me he tortured cats. And you date it from the time of Brian's disappearance? Didn't any of you question where he'd gone?'

'Of course we did, Kel, but what could we do when even the police weren't interested?'

'My dad? You can say it.'

'So you know all about it then?'

She nodded.

'When Michelle's old fella threatened Brian and your old fella did nothing about it, we knew to stay low. Brian didn't. He would never be told what to do. He had this warped sense of justice and a need to protect Michelle, and get revenge.'

'Revenge for her dad beating her up? I'd say that was noble.'

'You would say that. But when even the police do nothing, you kind of lose faith in justice. We were just pains in the arses to them. No one wanted to rock the boat. It riled Brian.'

'Were you aware that Mr Thompson – our geography teacher – was involved in some illegal trade with Jason?'

'Is that what you call it? Illegal trade?' He smiled. 'Yeah, we all knew. You were Miss Goody Two Shoes, weren't you? Were you shocked? I heard the old fucker was still working. People bang on about weirdos in schools, is there any wonder why? When my girl goes there, I'll be watching him like a hawk.'

'He's retiring this year.'

'Thank God. Should have been sacked years ago.'

'Carol had a few choice words to say about him too. So it's not all hearsay, then? You'd go on record and say that Mr Thompson was bent?'

'Oh come on, Kel. You really didn't see it? You always had your head in a book. It's why Dave fell in love with you. You were untouchable. Exotic.'

'Weird.'

'No. Not weird. Different. Better.'

'Dave didn't fall in love with me, he's not capable of it. The last time I saw him, before I banged him up, he tried to throttle me, remember? You were in court, you heard it.'

'Yeah. We went to school with some pretty messed up people, eh?'

'That night, when Dave drove you all to Derwent, and Michelle and Brian wandered off to Calfclose Bay, do you know what they talked about?'

Paul shook his head.

'They weren't talking when they came back to the car. I remember that.'

'And what happened when you got back to Penrith?'

'Like I told the coppers – your dad – I walked Tracey and Carol home, well actually, carried them.'

'I believe that Brian and Michelle hooked up with Jason,' Kelly said.

'But Jason wasn't out,' Paul said calmly.

'He was later, according to Michelle. And Mr Thompson was driving Jason's car.'

'You're kidding me?'

She looked into his eyes and realised that she had no idea what he was thinking. She'd lost her ability to read these people, because she knew their eyes. She'd shared their spaces before and her brain was wired to look up to them, in awe of their confidence and self-possession. She'd learnt to shut out their capability for deceit years ago because she'd never seen it.

She needed a break. All she had was stories going round her head, with everybody telling her something slightly different. And she had no solid evidence with which to challenge them.

'In your original statement you said you and Brian had seen Michelle go into her house.'

'Did I?'

'Yes.'

'Well I must have, then. Maybe she went back out after? You know her parents weren't exactly the most careful when it came to Michelle's freedom, she could pretty much do what she wanted, because they were always hammered. She hated Ken, she would have told him nothing.'

'I know.'

'Can't you just let it go, Kel?'

'What? Why would I do that? Brian deserves justice. I can't bear the thought of him at the bottom of that lake for twenty-four years, with whoever killed him still walking the streets.'

'What if they're not? What if it was really a random case of him being unlucky? If he was that shit-faced, then maybe they got split up and he wandered down a road in the dark on his own, got run over, and whoever it was panicked and threw him into the reservoir.'

She looked at him.

Melancholy overcame her. He was partly right, but not in the way he thought. She knew from Brian's injuries, even though he was just a skeleton, that Henry had had found no evidence of accidental death, such as being run over by a car, or falling.

'It's indisputable that he was murdered, Paul.'

'But you don't have a suspect, do you?'

She wanted to be mad at him, and she desperately hoped that he wasn't toying with her, but the warmth in his eyes gave her the feeling that she should let go. Thousands of cases every year went unsolved because the police simply couldn't work out what happened with enough corroborative evidence. Twenty-four years was an interminable period when you needed forensic proof to bring someone down. It was almost pointless. But she questioned why Paul would want it to be.

'How well did you know Ken?'

'Not really, he was an arsehole.'

'Was he capable of following through on his threat to Brian?'

'Of course he was. He was a violent bastard. Anyone with intent could kill a drunk teenager, couldn't they? But he was always pissed. How could he have got Brian to Thirlmere? He wasn't that smart. If it was Michelle's dad, then he'd have left him where he was.'

'Good point.' Kelly sighed.

'Fancy a pint?' Paul asked.

Kelly smiled. 'I'd love to but I'm a mum now, I need to get home.'

'Maybe some other time?'

She nodded and got up to leave. 'If you remember anything else will you give me a call? Do you see any of the others anymore?'

'No. I grew out of driving around all night looking for a place to smoke weed. I got a job, I've got a mortgage. Now I've got a daughter too. I support her mam. Life gets more complicated doesn't it? And when Dave went down, I distanced myself further.'

'So, did Dave see them still, I mean before he went to prison?'

'You didn't know?'

'What?'

'The defence were going to call Steve Thompson as a character witness for Dave at his trial, but they dropped him because they decided what might come out would turn him into a hostile witness.'

'What might come out?'

'I might have come to his trial, Kel, but I hated what he did. He got what he deserved, I just thought you should know that.' He paused. 'Dave blackmailed Mr Thompson for years. He threatened to reveal his little money earners on the side if he didn't show him the ropes.'

'Show him the ropes?'

'You know, how to get stuff into the country, how to avoid the police, who to sell to, how to manipulate people, all that. He told me. Steve Thompson was always doing something illegal to make money. I guess teachers' wages are really shit. But he was small fry. Anyway, it was considered too risky for Steve to be put on the stand under oath, Dave was going down anyway. He was desperate, Kel, you have to remember that.'

'I do, I remember his hands around my throat.'

'Why didn't you know?'

'The defence don't have to reveal their sources, and only have to inform the prosecution of witnesses they actually use. We don't see the ins and outs of them constructing their case. Steve Thompson was never mentioned. I always thought he was a nice bloke.'

She felt slightly ill. It was the familiar feeling of being duped. She hated being taken for a fool. She felt her cheeks go pink, but it was hot in the small garden.

'He had everybody fooled didn't he? Anyway, I just thought you should know, because if he was driving Jason's car then maybe he knew more about Brian's disappearance than he's letting on?'

'Do you think Dave had more on him than just his dodgy transactions out of school?'

'Put it this way, I reckon Thompson was relieved when Dave went down.'

'But he could still blackmail him from prison,' Kelly said. 'Jesus, Paul, is there anything else?'

Paul smiled. 'That drink sometime?'

'Sure. Thanks for the tea. When are you seeing your daughter next?'

'This weekend.'

'Treasure it,' Kelly said. 'My other question is, when was the last time you saw Jason?'

He shrugged. 'Maybe a few months ago, drunk in The Cock, in Keswick. Spouting off about how he could drop some fella. I left as soon as I saw him.'

Kelly nodded and thanked him for his time.

Paul showed her out and she walked back to her car with a feeling of frustration. She remembered Dave's smug face when she visited Highton prison this morning with Rob, and decided to take a look at what Dave Crawley was being sent in prison by way of comforts, and who from.

Chapter 37

The end of the day couldn't come quick enough. Kelly's head throbbed with information and she knew that her body and mind needed a break. All she wanted to do was crawl into bed, after a good meal, and sleep all night without Lizzie waking her up. But she was starving, and when she returned home, after driving five minutes around the corner from Paul's, the Chinese had arrived and Johnny was setting the table. The smell only made her more ravenous.

'God, that smells good.'

It was a full house and the noise filled her head so there was no space left for dead bodies, lying school friends, con artists behind bars, and bent teachers, or coppers.

'You look tired,' Ted said, giving her a squeeze.

The cartons of food were arranged on the kitchen table and Josie got out plates and cutlery. They each took their turn piling their plates with food and slopping sauces over the top. Lizzie bashed into cupboards in her walker and Kelly cut up her food in a bowl, ready to give her mouthfuls of flavour bombs and watch her reaction. She'd read that a baby's first battleground was food: they sniff the power they have over you when you make it a test of will, and so Kelly insisted that she take what she wanted, and no more.

She took off her shoes and sat on the floor. Lizzie pushed herself over to her with her mouth open, and Kelly gave her a spoonful of chicken with sweet sauce. The child bent her knees vigorously and flapped her arms about.

'I think she likes it,' Henry said.

The others sat down with their plates on their knees and Kelly felt a slight breeze coming in from the open terrace doors. In between feeding Lizzie, she stole spoonfuls of food from her own plate which she placed on the floor beside her.

'I passed my instructor's course,' Johnny said, when they were all settled and eating happily.

'Oh, sorry! How did it go?' Kelly asked. She'd completely forgotten that Johnny was being examined today on his skills as a gorge walking and abseiling instructor.

'Well, I can now take out groups of up to ten. I did the written exam this morning and then we went up Glenridding Beck after that. There wasn't much water, obviously, but there was enough. I'll need to do a winter refresher because the conditions are so different.'

Lizzie bumped against Kelly, demanding more food, and she obliged by putting a full spoon of rice into her open mouth. Johnny's desire to become an adventure guide had been something he'd been toying with for a good few years, but had become serious about recently. He had plenty of mountaineering and first aid qualifications, which were necessary for his mountain rescue work. It made sense. She couldn't imagine him sitting behind a desk. Even his PTSD counselling had been conducted outside, on walks, or doing some activity. He belonged outdoors, using his body. And he was good with people. It eased their rhythm; the outdoors soothed him.

Kelly thought about how she might contribute to the conversation and tell everybody about her day, but she couldn't think of anything to say that didn't involve murder, abuse and bad people. She looked at Ted, who she could tell was thinking the same thing. It was all right loving one's job, but sharing it had to be done sensitively.

Henry launched in regardless, believing that everybody shared his fascination with the dead, and what their remains could tell him.

'I once had a case of a gorge walker going missing here in the Lakes,' he said.

'Gosh, I remember that,' Ted said.

'She couldn't swim but the group leader didn't check. She jumped in a pool at the bottom of a rather large waterfall, thinking it was shallower – it was called something Ghyll.'

'Esk?' Johnny and Kelly said at the same time.

'Yes, that was it.'

'And she couldn't swim?' Johnny said, shaking his head.

'It was terrible, they pulled her out further downstream, but she was already dead.'

Kelly looked at Josie, who was engrossed. She glanced at Kelly wide-eyed.

'I don't remember that,' Kelly said.

'We're going some years back now.'

'Did you examine her, Granddad?' Josie asked.

'I did. I was the one who called Henry because I knew he'd want to get involved.'

'Anyone would think she'd been run over by a steamroller,' Henry said.

'I thought your field was skeletons?' Josie asked, ever the stickler for detail.

'You're right, my dear girl, but I examine what happens to bones in a good variety of situations.'

Kelly paused with a spoonful of food for Lizzie, but shovelled it in anyway, knowing that now Henry was on a roll, and Josie a more than willing audience, he wouldn't stop until he was finished. Johnny winked at her.

'Your grandfather called me because of the X-rays. He knew it'd be a fantastic case for me to study, if the family agreed.'

'And did they?' Josie asked.

'Yes, they were kind enough to donate her body to medical science. They were atheists, you see, so they didn't believe in the theory that your body after death is sacred and needs to be treated in a certain way.'

'Did you have something to do with talking them into that, Henry?' Kelly asked.

'Well, actually, I had a very candid conversation with her father. He believed that her spirit was set free and so what was left was a mere shell. However, we of course treated her with utmost respect.'

'I heard that medical students throw body parts around mortuaries,' Josie said.

Ted almost choked on his pork rib.

'I think you'll find when you get to university that it's not just medical students who are badly behaved, Josie,' Henry said.

She laughed, and went to fill up her plate for a second time.

'Oh, I needed this,' Kelly said.

'You want me to take over?' Johnny asked.

'No, I haven't seen her all day. I don't mind. I'm getting one mouthful for every two of hers, it's fine.' She smiled and grabbed another spoon of noodles.

'She's got an excellent mind, Johnny,' Henry said, referring to Josie.

'Well she doesn't get her brains from me, I can tell you.'

Lizzie stopped taking the spoon offered to her by her mother so readily, and Kelly wiped her mouth as best she could before she darted off across the room. Kelly sat against a sofa and finished the rest of her own food.

'How is it going, Henry? Did you have a productive day?' Kelly asked.

'Well, we did as a matter of fact. We've got plenty to discuss. I don't want to speak out of turn in front of the young lady,' he added.

'What don't you want to talk about in front of me?' Josie said, returning with another plateful of food.

'Work,' Kelly said.

'Oh, please, it's so much better than Netflix,' Josie said.

'I'm sure you'd get bored eventually,' Kelly said.

'No, I won't. Have you caught anyone yet?' she asked.

'No. It's been two days!'

'I meant the body in the lake,' Josie said.

'Ah, of course. Well, that's Henry's field.'

'May I?' Henry asked.

Kelly looked at Johnny, who shrugged. 'She's old enough.'

'No names,' Kelly said.

'Well, you know there is a huge amount we can learn from bones,' Henry began.

Josie was mesmerised.

Kelly reckoned that Henry must have grandkids, because the way he spoke to Josie was bang on for her age. He stuck to stories of bones and examinations and all the things they could do in the lab to piece together somebody's life, and death.

Kelly finished her food and felt herself nodding off.

'Have you ever lectured, Henry?' she asked him.

'Yes. From time to time, but it's not a wage you can survive on, is it Ted?'

They agreed. Kelly reckoned that the more interesting a person's job, the less they actually got paid for it, but that was the way of the world.

Johnny cleared up. Kelly felt as though she couldn't move but she knew that she couldn't stay on the floor all night. She dragged herself to her feet and helped clear the pots and cartons away. There was little left but Josie still came looking for another serving.

'I went to the gym today,' she said. 'God, he is so interesting!'

'Isn't he? It's amazing what he does, and your grandad,' Kelly said.

Josie went back to the lounge to ask more questions.

Kelly would have her chance to quiz Ted and Henry later. But for now, she enjoyed the hum of her family all talking over one another, coming and going, picking up the pieces that fell off her during the day. She had a reputation for having Teflon shoulders. But it wasn't true. The things she saw left an indelible footprint. To many, a Chinese takeaway and endless chatter might seem ordinary and even quite annoying, especially if somebody finished the ribs without telling anyone else. But to

Kelly, it was ointment for her wounds. Each time she absorbed the impact of a murder – of seeing somebody who was once alive and vibrant, like her own family, on the mortuary slab; of seeing the pain and mess they left behind – it carved off a little piece of her.

Chapter 38

Josie was up and out early to meet a friend, to take the train to Kendal. They were going to the cinema and having lunch out, as well as doing a bit of shopping. The film was a matinee showing of a classic horror and Kelly often wondered if the girl's mind would be better suited to a forensic science degree, rather than history.

Johnny had a free day after his taxing efforts yesterday, and planned to take Lizzie out.

'Where are you off to?' Kelly asked, noticing him getting the carrier ready, plus all sorts of outdoor gear.

'We're taking the steamer to Glenridding, then I'm going to walk back along the lake path,' he told her.

'Pack sun cream,' she said.

'I know. She's got a hat, and I've got plenty of water. I'm in no rush,' he said, kissing her as he rushed around, gathering all the detritus that one needed for an excursion with an eleven-month-old.

'We might go for a swim,' he said.

Kelly ate toast and sipped tea, and reflected how hard he was working on keeping their family together. Their split last year was never discussed. There was no need. They were either going to make it work or not. Ted and Henry had spent the night at Johnny's flat and had arrived at her house early to take her through their findings from yesterday. They sat at the kitchen table.

'So, the model is ready,' Henry said jovially.

He swung his laptop around on the table so Kelly and Ted could see it.

'All the information is fed in, and the software comes up with an animated film of what actually happened to the victim.'

'Impressive,' said Kelly, chomping.

'Isn't it just?' Henry said.

'This is Brian,' Henry said.

Kelly saw a skeleton on the screen and she remembered the face of the teenager who'd been reduced to a computer file.

She saw a human form in the standing position, then a restraint was applied to his wrists and he was laid on his side. She was imagining the back of a car boot, perhaps.

'The force required to produce such trauma to the skull would need a good back swing and so it's likely he was on the ground when it was done.'

Kelly envisioned Brian being taken out of a boot space by force. For some reason, she saw his yellow socks in her head, and it made her even sadder.

An animated figure appeared on the screen and swung back what looked like a spade, and brought it down with force. A close-up then appeared and showed the damage to the back of the model's skull.

'The pattern matches perfectly,' Henry said. 'The surface area, times the force, gives us the exact impact wound that I've seen on Brian's head. It would be flat, metal and about twenty to thirty centimetres across.'

'Like a garden spade or a bit of wood?' Kelly asked.

'Wood isn't dense enough, we tried it with the weights and dimensions of a piece of heavy timber. It would absorb some of the shock.'

'What about the human variable? We don't know who swung the weapon,' Kelly said.

'We input three variables, an average woman, man and adolescent.'

'And what was the result?'

'An adult man,' Henry said.

She sat back. 'How accurate is the science?'

'I've used it in dozens of cases,' Henry said.

'It's superb. I've been involved in several cases where it's swung a jury,' Ted said.

'Now for Jason,' Henry said.

Kelly saw the same computer-aided short film, but this time, the human had flesh on his bones. Kelly sensed Ted's excitement and peered up at him. He smiled at her and nodded at the screen. He'd clearly already seen the animation, several times.

She watched the projected timeline of events for Jason's injuries, and what caused them, and then sat back, horrified.

'A spade?' she asked.

'The head wound profile fits, with four different strength profiles of the wielder of the weapon,' Henry said.

'But it's the neck wound that convinced me,' Ted added. 'Look.'

The screen switched to a close-up of the cervical spine of the victim. Kelly kept having to remind herself that the cartoon-like character was a real human. It was like watching a sci-fi sequence from a movie, but she knew that it was no fiction. An edged weapon was shown penetrating the neck at an angle, and tearing the soft tissue as far down as the spine. It then smashed the disc and the cartilage in between three vertebrae as well as causing catastrophic damage to the bone itself. She watched as the weapon was withdrawn from the wound and the edge of it, as well as its dimensions, appeared in another animated slide.

'We spent hours yesterday running all the heavy weapons we could think of into the system, and this was the only one that could do the same amount of damage, with the exact trauma profile,' Ted said.

'So, it's pretty certain that Jason was surprised from behind and never even saw his attacker?'

Ted nodded.

'He was stunned with the first blow, though it would have eventually killed him. He would have been knocked unconscious fairly quickly, but then he was struck with the edge of the spade when he was on the ground, with the intention of near decapitation,' Ted said.

'Premeditated and calm,' Kelly said.

'And full of rage,' Henry said.

'They took their time to turn the spade around to inflict the second wound,' Ted agreed.

'It was an act of obliteration, and no struggle,' Kelly said.

'Corroborated by the absence of defence wounds,' Ted said.

'Pretty conclusive.'

'It also lends weight to the theory that whoever killed Brian, killed Jason too,' Ted reminded her.

'I think that's going to be impossible to prove, but I will say that if I bring Jason's killer down then I will feel satisfied that Brian has some justice too. Unless it was set up like that, and it was Jason who killed Brian all along.'

'Really? You had a busy day too then yesterday?' Ted asked.

'Yes, lots of people telling me lies and obfuscating the truth. Many of the people who could help me with my inquiries are deceased, and it seems increasingly unlikely I'll solve Brian's case,' she said.

Ted touched her shoulder. 'Well, from a scientific point of view, you have two brutally chilling murders with much in common, including the type of murder weapon. I'd say that if you get close to solving one, the other will follow for you,' Ted said.

She knew he said this because he loved her and wanted her to feel satisfied and vindicated in her work, just like he was, and indeed Henry too. If only it were that simple, she thought. If she could input her suspects into some computer software and press play, and the algorithm was one hundred per cent foolproof and reliable, and at the end it spat out the guilty party, how that would change crime stats, she thought. Maybe that's why Ted and Henry preferred working in mortuaries, because they didn't have to face liars who hid things from them.

'Well, gentlemen, it's been a pleasure. Please help yourself to bagels, pancakes, and whatever you fancy. I've got a lake to search. You never know, I might get lucky and find a spade, rather than just some bird shit and old tin cans. Call me cynical.'

'Happy spade hunting,' Henry said.

They heard Lizzie squeal and knew that Johnny had got to the stage in his packing where she'd worked out that they were going out. Like showing a puppy a lead and a poo bag.

'I'm so jealous, have a fantastic trip.' She kissed both of them, then hugged Ted and Henry.

Then she rushed out of the door and got into her car and pulled away, driving past the Chestnut House and noticing, for the first time in all the years she'd lived there, Paul Gordon buying coffee.

Chapter 39

The decision to search the lake at Parkie's for evidence wasn't taken lightly. It wasn't that the lake was deep, particularly dangerous, or impossible to chart with sonar, it was more that the odds of finding anything were slim. They were looking for a potential weapon that was no bigger than a large canine police dog, in the space roughly the size of a football pitch, and five metres deep in the centre. There were reeds, thick mud and submerged aquatic flora that the divers would have to negotiate. Michelle had informed them that Japanese Pondweed had taken a hold too. Kelly parked next to the office and walked to the lake, spotting Michelle, who was making brews for the team. Kelly could see that a RIB had already been launched into the water from its trailer, and it was manned by three divers. A large truck was parked as close to the water's edge as the silt and mud would allow. It was used to collate data and communicate with the divers, as well as track progress. Kelly could watch the live feed from their body cameras.

'Ma'am,' the sergeant of the Underwater Search Unit greeted her with a warm handshake. They'd worked together before, with Johnny too. If there was one thing in abundance in the Lake District that gave ample opportunity to conceal illegal kit, it was water. The mountain rescue often called in help from the underwater unit when some unfortunate soul slipped off rocks into the depths below, and was either swept away, or simply couldn't swim.

'Keith, how are you this fine morning?' she asked.

'Not bad. Straightforward search, is it?'

Kelly nodded. 'We're looking for evidence, so anything unusual that doesn't belong in a lake.'

'My favourite.'

'Yeah, not a body, right?'

He nodded.

'What's the plan?' She deferred to his expertise.

'We're going to follow a grid pattern and search each square at a time. We might need more than one day,' he said.

'Sure.'

'Come on, I'll show you. Do you want to sit in the van and keep an eye on us?' he asked.

They walked towards Michelle, and Kelly said good morning. Thelma and Louise sat panting beside her and smiled up at Kelly. She petted them.

'Thanks for agreeing to this, Michelle.' Kelly tried to act warmly, giving nothing away.

'I'm just hoping it doesn't disturb the wildlife too much,' Michelle said. 'I've got nesting sand martin over there.' She pointed in the distance to the opposite bank of the lake, where there was a reed bed. 'They're protected by the Wildlife and Countryside Act.'

'We'll try our best,' the sergeant said. 'It's not invasive, and the boat is quiet, and we're underwater, not on the surface,' he added. 'We're also concentrating on this side of the lake.'

Michelle smiled. 'Coffee?' she asked Kelly.

'That'd be great, I'm going to go down to the lake,' Kelly replied.

'I'll bring it down for you,' Michelle said.

Kelly went with the sergeant to the large truck and stepped inside. It was kitted out with computer screens and bleeping machinery. She sat on the chair pulled out for her and watched as the divers on the boat made themselves ready. She heard their preparation and was familiar enough with most of their jargon to know what they were talking about. She knew the routine of checking their equipment, and their air, and watched as they plopped in over the side of the RIB.

She watched the monitors linked to the chest-mounted cameras carried by the divers and saw a murky soup of green and brown. The radios crackled and the sergeant in charge concentrated on his task. An eerie silence descended on the van and Kelly tried to make out what the divers were looking at. They'd started close to the water's edge, working on the theory that anyone lobbing a spade into the lake could have only got it so far.

Michelle came in behind them with two mugs of coffee. One for Kelly and one for herself.

'What are they looking for?' Michelle asked.

'Anything that could be used as evidence. You never know, Jason's murderer could have dumped something in here. It's quite common for killers to panic,' Kelly said.

Michelle watched the screens.

'What would the protected birds do if they were disturbed?' Kelly asked.

'They'd desert the nests, I guess,' Michelle said.

'And they've seemed happy enough this last week?' Kelly asked her.

'Seems so. I check them every day.'

'So, you'd notice if their nest was destroyed? Or if they were spooked by something?' Kelly asked.

'Of course. They'd be either rebuilding frantically or they'd have gone.'

'Maybe follow this pattern,' Kelly said to the sergeant, pointing at the grid.

He nodded and communicated to the three divers, who were making good progress, but finding nothing of consequence. Kelly knew that anything thrown into the lake on Tuesday afternoon or evening would be visible even in such gloomy conditions. An item such as a spade, or a heavy bat with a metal club for example, would have simply settled on the bed itself. It would take more than three days to cover over in silt and mud. She watched as the divers went deeper; though the lake itself wasn't a challenging dive, it was tricky enough because they had to be mindful of their body positioning so as not to disturb too much detritus from the bottom.

She knew from experience that watching lake searches, or searches of any body of water for that matter, was a tedious task and could take all day with no results. Her plan was to leave the team to it and check in from Eden House throughout the day.

Michelle glanced at her, seemingly mesmerised by what was going on under the water.

'It's beautiful,' she said.

Kelly looked at the screen and had to agree. The cameras picked up the penetration of the sunlight through the depths and the colours changed rapidly as the odd cloud crossed the sky above. Different shades of green shimmered in lines, like rainfall, and the reeds floating upwards made it look like a scene from an underwater movie. Five metres wasn't deep, but the divers still had to pay attention to every contour and shape they saw, at times feeling around with thick gloved hands. The hazards inherent to the job were not necessarily drowning or losing one's way, as in recreational diving, but the objects and sharp edges they might encounter in waterways. People threw all sorts of rubbish into the lakes around the national park.

'I wanted to ask you,' Kelly said to Michelle. 'Did you provide a storage facility of some kind for Jason? It's been suggested that he traded in a fair amount of farm goods,' Kelly asked her.

'Stolen, you mean?'

'I can neither deny nor confirm that,' Kelly said.

Michelle rolled her eyes. 'Of course you can't,' she smiled. 'Well, he had keys to a shed, if that's what you mean. Sometimes customers want to store water sports kit and the like, so I provide a few units for them. Jason had one.'

Kelly could kick herself for not asking before. It was only when she read the forensic report on Jason's cabin that she realised that the accusation from Connor's farm didn't make sense. If Jason was a man of limited means, and he'd been living hand to mouth, courtesy of Michelle's generosity, then he'd have to have somewhere to store his stuff, and his car was empty.

'Can you show me?' Kelly asked.

Michelle agreed and they left the truck, carrying their coffee cups, wishing Keith good luck. Michelle grabbed some keys from the office, then they walked to a shaded area, and Kelly spotted some wooden box sheds under a clump of trees.

'You wouldn't even know they were there,' Kelly said.

'That's the point, I want the site to be pretty,' Michelle said. She unlocked the one that she said was used by Jason, and Kelly peered inside. It was full of tools and bags of items. Several spades, forks and other heavy equipment was propped up against one side. She didn't touch anything but peered closely at the collection of spades, which she knew could potentially be used as lethal weapons. They appeared clean, and brand new. Nevertheless, she'd call forensics back to itemise everything and look for any evidence that somebody other than Jason might have been inside.

'Who else had access to this apart from you and Jason?'

'Nobody I can think of.'

'Have you any idea where his keys might be?'

'I assumed he always had them with him, or inside his cabin, maybe his car?' Michelle said.

'Thanks, Michelle. Can you lock this up and make sure no one goes in until I've got a forensic team back here?'

Michelle did as she was asked and popped the keys in her pocket.

'Right, I'm going to check on the divers and then head off. I heard from the coroner that Brian's body will be released back to his family. Will you go to his funeral?' Kelly asked.

Kelly had already spoken to Maureen and Donald Miller, who'd told her they'd want a proper burial when Brian's body was released. She'd have to go. It wasn't just because she'd gone to school with him, and his disappearance had rocked the community, but also because funerals were good opportunities to watch suspects. Many killers couldn't keep away from their accomplishments. Some popped up on TV desperate to be interviewed for their five minutes of fame, others revisited the site of the kill, and plenty attended funerals to take some kind of vicarious pleasure in the sorrow of the loved ones.

'Of course I will. I tried to get in touch with Jason's sister to ask her about it too, but she's not answering her phone.'

'How did you manage to get her number?' Kelly asked. Joanne Cooper had been hell to track down.

'I had her in my phone. She was my last resort when Jason ended up in trouble,' Michelle said.

'I thought she washed her hands of him years ago,' Kelly said.

'She tells people she did, but she still loved him.'

Chapter 40

When Kelly got back to Eden House, Danilo Alves was waiting in the foyer. Kelly recognised him. He stood up when he saw her, also recognising her from the farm. He was cleanly dressed and shaven, and Kelly was struck by how less aggressive he appeared out of his farmhand garb, and she felt a pang of guilt for judging him the other day when she'd visited Cooper's.

They shook hands. 'Mr Alves, would you like to follow me? Thank you for coming in. I'll take you into a private suite, where we can chat properly,' Kelly said, putting him at ease.

She took him into an empty interview room and sat down, indicating for him to do the same. He was clearly nervous and Kelly noticed him picking at his hands and glancing about. Coming into a police station was stressful for anybody and she took it as an opportunity to watch him closely.

'I'd like to ask you a few questions about Jason Cooper, who you worked with at Connor's.'

He nodded.

'It's an informal interview, you can relax,' she added.

She saw a shadow of a smile and felt some sympathy for him. His alibi had checked out for Tuesday. Three fellow workers had confirmed he'd gone to his caravan after work, and they'd watched Netflix for the rest of the evening. Jason Cooper's name had come up, in that they'd discussed what a dickhead he was, but that was it.

'So, I want to draw your attention to what you told us about Jason's alleged illegal dealings in tools from the farm. Kevin was good enough to forward me a list of equipment unaccounted for. It would seem that somebody, not necessarily Jason, was on the take. Let's see.'

She brought up Kevin's email, which she'd received only this morning, and browsed through it.

'This is quite expensive kit.' She turned around her Toughpad and showed Danilo the list. 'What made you suspect Jason?' she asked. She thought of the brand new tools in the shed that Michelle had shown her and knew that at least some of it matched the list. She'd have to check thoroughly, because her knowledge of farm tools was rusty at best, though she knew what a spade looked like, and a brand new one at that.

'He didn't try to cover it up,' Danilo said. 'He bragged about it when he was drunk.'

'Was he drunk on the job?'

Danilo nodded. 'That's why we got into a fight. He was a liability. That's why he was going to be let go.'

'You know a lot about the Cooper's business decisions for a labourer,' she said.

'Kevin trusted me to get more involved. It was me who told him about the missing equipment. I saw Jason loading his car with it.'

'So, you caught him red-handed?'

'What is "red-handed"?' Danilo asked.

'Sorry, it means that he wasn't even trying to hide it, like he was caught in the act, as it were. It's a Scottish saying.'

'Not English?'

'No, I was told off for suggesting it by a colleague of mine who's Scottish. We're very protective of our sayings,' Kelly said.

Danilo visibly unwound.

'So, when you fought, was it something that had been building up or was it out of the blue? Sorry, there I go again. I mean, a surprise?'

'I don't like people who cheat. The Connors are good people. Jason caused them many headaches. I'm not a violent person, but I snapped.'

'It happens. Was this the only time? Did anyone else find themselves in this situation with Jason?'

'He wasn't liked. In my country we say, *Diga-me com quem andas e eu te direi quem tu és*. It means "tell me who you walk with and I'll tell you who you are".'

'As in he kept bad company and it was obvious?'

Kelly could have listened to Danilo speaking his native tongue all morning; there was something soothing about listening to sentences one didn't understand, and didn't have to.

Danilo nodded.

'Do you know who he kept company with?'

'No. That's the point. I didn't want to. Occasionally, I'd see him drinking in Keswick – it's a small place, for somebody who is looking to explore at least – and he was the kind of person who sat at the bar alone, telling stories. The type of person you avoid. Why do the English drink so much?' Danilo asked.

Kelly reckoned that if she understood the endemic national obsession with alcohol abuse then she'd crack many cases before they happened, a bit like the film *Minority Report*, but without the need for precogs.

'I think it's something to do with the breakdown of the family unit, but you could get a thousand answers to that question. It would certainly make my job easier if we didn't rely on it so much,' she said. She was a little ashamed of her nation's reputation for booze. It was made more pertinent by the amount of people in Jason's life blaming the poison for the mess he'd made of his life.

'So, Danilo, I've got some photographs of equipment found in a storage facility, and I want you to have a look at them for me.'

She showed him the photographs of the tools she'd seen at Michelle's in the small shed. They were stacked up and clearly brand new, which gave her the impression that Jason had no intention of using them. Danilo flicked through them.

'They're definitely the same type used at Connor's. I order them so I should know.'

Kelly noted the level of trust the man had established with the farm.

'We use this brand all the time. They're supplied by Spaldings, in Lincoln, and they only use Bulldog. Look, you can see the label.'

Kelly looked and saw the brand name clearly.

'Can individuals order from them or is it trade only?'

'Both. Trade carries a hefty discount though and you need to sign in. Did Jason have a phone or computer with orders on it?'

Kelly reckoned Danilo was suited to detective work.

'We're working on it.' She'd entrusted Rob to that particular task, after the forensic team had finished inside Jason's cabin. He'd had a laptop, and his phone was found in the pocket of the shorts he was wearing when he died. Typically, these items were sent to sterile labs to open and decode but it took time to get the results back.

'If there's no evidence that he bought them, then it indicates they were stolen, yes? Besides, if you can get me the barcodes, which should be on the back of each tool, I can check if they were ordered by Connor's.'

'Perfect, thank you. Do you recognise this man?'

Kelly showed him a photo of Steve Thompson. Michelle and Paul had both said that Jason drank with the teacher from time to time. It would seem that their relationship went back years, but Kelly still didn't understand it.

'He looks familiar, but I can't remember why,' Danilo said.

It was better than nothing. 'Take your time. Have a think.'

Danilo took the photo from her.

'I'm sure he's been to the farm,' Danilo said. 'I think he dropped Jason off a few times. Jason often looked, what do you say? Worse for wear? I think he could not drive sometimes.'

'Why do you think?'

'Alcohol? He was sent home on a few occasions by Kevin, for not being able to work. And this man picked him up, I'm sure.'

Kelly brought up the Westmorland Comprehensive website and showed Danilo a different photo of Steve Thompson.

'Him?'

'Yes, that's him. I thought it might be his father.'

'So you say Jason transported the stolen goods from the farm in his car? It was that easy?' Kelly asked.

'He thought we wouldn't snitch.'

'But you did?'

Danilo nodded.

'But you said that equipment was returned after you'd complained?'

'I think it was a one-off, to save his job. He'd been caught, like you say, red-handed, and the stuff reappeared as if by magic on Tuesday afternoon.'

'Everything?'

'No. Not everything. If you're showing me photos of shovels and spades that were in his possession then I guess he kept some, no?'

'So, you're suggesting he brought some kit back, but just to make it look as though he wasn't in the wrong, he didn't return everything?'

Danilo nodded again.

'And did you ever see this man helping Jason with the tools?' Kelly asked, pointing at Steve Thompson.

Danilo shook his head. 'But I heard Jason on the phone arguing on Tuesday morning.'

'Did you hear enough of the conversation to understand what it was about?'

'He was saying that he was – excuse me – fucked up. No, that *it* was fucked up. It was all over.'

'What was all over?'

'I don't know, he just said that he needed help. Whoever it was he was talking to seemed to calm him down.'

'Can you think about that conversation carefully and give me any indication at all what it was about?'

'Let me think. He said that it was all over. He was talking in the barn. I went in there to change the oil in one of our tractors.'

'Did he know you were there?'

'No.'

'Do you know what time it was?'

'About ten, because I'd already been out to the field and come back, and it was almost brew time.'

'Anything else?'

At this point, Kelly watched him closely. His body was giving her no cause to suspect what he was telling her was anything but the truth. He was open, and his face sincere. She looked for signs of stress and deceit: leg tapping, head shaking, undue earnestness, narrative leading and the like. There was none.

'I can't do it, Steve,' Danilo said.

'What?'

'He said, "I can't do it, Steve."'

Kelly knew she hadn't told Danilo Alves the name of the man in the photo.

Chapter 41

Kelly greeted the rest of her team upstairs. It was still early but the incident room was buzzing with activity. She could see Rob speaking to a small group near the window, and Kate was writing on the whiteboard. Tomorrow was the weekend, but Kelly knew that she'd have her hands full. They were waiting on results from several labs, and until then they had to sit tight and follow the leads they had so far.

One of those was the woman who'd said she'd seen Jason's red Ford Fiesta near Thirlmere on the twentieth of July 1997. Kelly thanked the habits of local people who rarely moved very far, and saw that the woman still resided at the same address. She checked the phone number and called it. She introduced herself to the woman who answered and it didn't take long to establish that the woman remembered the call as if it was yesterday. Twenty minutes later, Kelly was still on the phone, as the woman had poured out her fears and sorrow over the recent findings at the reservoir.

'If they'd listened to me he could have been found years ago,' she said sadly to Kelly.

The woman was distressed. Rightly so.

'Can you remember who you spoke to back then?' Kelly asked.

'Of course I can, it was John Porter. He said he'd look into it and I assumed he did, but now that poor boy was found over here, I don't think it was taken seriously, do you?'

'I'm sure it was,' Kelly tried to reassure her. 'The incidents might not have been connected,' Kelly said. She closed her eyes as if that would make the difficult conversation go away. 'When you say over here, I see you live near the reservoir. Would it be all right to send an officer to check your statement? I know it was a long time ago.'

'It was like yesterday to me, because the water company was doing maintenance on the reservoir and we had all sorts of vehicles up and down all day and night for months. But this wasn't a water company vehicle. It was swerving all over the place. I got the number plate. We don't get things like that happening round here, it's super quiet. It was more than unusual.'

Kelly closed her eyes as the woman fiddled with paper and read a number plate to her. She didn't need to write it down. It was Jason's car.

Kelly managed to talk the woman into holding off until she could send an officer to her house, but inside she felt as though she'd betrayed her public role. She felt responsible for whatever had transpired in this very office twenty-four years ago, and she felt miserable. She got the woman's address and went to Dan, who sat working at his desk.

'Can you pay a visit to a witness on the Brian Miller case?' she asked him.

He swung back on his chair and nodded.

'I could do with a break,' he said.

She explained who the woman was and what she'd said.

'I can get on it as soon as I've sent a few emails. By the way, Steve Thompson is coming in at eleven,' he said.

'Great.'

'And I've found the red Ford Fiesta. It belongs to a keen collector. But he lives in Aberdeen.'

'Not so great.' Kelly knew that the chances of finding any evidence inside the vehicle after so long, with new owners thrown in, was virtually zero. And besides, any good barrister would argue that Brian, as Jason's friend, would be expected to travel inside his car. But they could check the boot... There didn't exist a hard and fast answer for how long DNA survived on surfaces, it all depended upon the conditions. In ice, for example, it could last millions of years, out in the sun, less than twelve hours. It was worth a shot.

'How are you doing?' Kelly asked quietly.

'I'm all right, boss. I found a place to stay,' he said.

She tapped his shoulder. 'Good. Let me know if you need anything. We're here to support you, Dan. I want you to know that,' she said.

'Aye, I do, boss. You've got a right pair on your hands haven't you?' he said, nodding at Rob. 'I offered him the spare bed at my new place and he's said yes.'

'Great, I can't imagine what a state it will be with you two living together. Do you need help moving stuff?' she asked.

'I'll let you know, boss. Have you got that address?' he asked. She gave it to him and explained that if he took the turning at Threlkeld onto the A591, he'd find the house at the northern tip of Thirlmere.

'There's a row of white houses on the left, and little else apart from a few hills leading up to Helvellyn.'

Kelly had hiked there plenty of times and often wondered who lived in the row of pretty stone houses, thinking them all holiday lets. The woman had told her that the red Ford had performed a screaming turn outside her house, and she'd peered out of the window, as the noise had woken her up.

'I'll get cracking,' he said. 'Any news from the lake search?'

'Not yet,' she said. She walked away and beckoned Kate to her office, who grabbed a pile of paperwork. Kelly hoped it contained good news.

Kelly sat down behind her desk and Kate sat opposite her.

'I'm going to see the superintendent. What mood's he in?' Kelly asked.

Kate smiled.

'That good?' Kelly asked. 'I doubt there's anything that can be done about the shitshow of an investigation into Brian's disappearance, but I need to get my concerns on record.'

'Well none of us were working here at the time so it's not exactly as if any of this is a mistake that happened under our watch,' Kate said.

'You're right. I've sent Dan to see the woman who lives near Thirlmere, who rang in with information about the red Ford back in 1997, on the actual night Brian was last seen.'

'I can tell you're angry,' Kate said.

'It's galling. First, I have an abusive domestic that's ignored, and now vital evidence being overlooked, seemingly on purpose. I always thought John Porter was a hero. I grew up listening to him telling me stories about the police and what a noble institution it was to work for.'

'As long as you didn't rock the boat,' Kate said.

'Or weren't a woman in need. I know the force was different then, but it was the Nineties for Christ's sake. We were still dinosaurs. If this comes out, the press will destroy us.'

'You need to get that across to Andrew. Maybe he'll have to bite the bullet and apologise to Brian's parents. There may even be a case for litigation. I'm no lawyer, but it's what I'd do, I think.'

'Would you?' Kelly asked.

'If it meant my son might still be alive? We know this much: Brian knew that Michelle's dad was abusive, and it wasn't taken seriously. We also know that her dad threatened Brian – a minor – with no consequence. Then he winds up dead, under the watch of John Porter, who didn't investigate properly. Have you thought about trying to trace some of his colleagues at the time?' Kate asked.

Kelly nodded. Then she looked at her watch. 'Can you interview Steve Thompson? Soften him up for me. Push him on his relationship with John Porter. I need a link. I need to prove that men in authority were doing nothing about kids who were vulnerable. Ask for his fingerprints, so we can compare them to any found in Jason's car, if there are any left. Somebody with his stature should agree to aid police – it would set alarms off if he didn't.'

'You know we still won't get hard evidence, even if you get somebody on record?' Kate said.

'See if you can get him to agree to a line-up as well, and get Mr Dougal in if he agrees. I can see the investigation into Brian's death going nowhere. And I can't get anybody on record who actually cared about Jason Cooper. It's as if he was snuffed out by somebody who knew he was so thoroughly disliked. Rob traced the red Ford.'

'That's great news,' Kate said.

'Not really. I can't see it being any use to us after twenty-four years.'

'Okay, so let's focus on what we have got, because I can tell you're in a really bad mood,' Kate said, tapping the pile of paper on her knee.

Kelly smiled weakly. It was true. Her frustration was affecting her ability to home in on what was important. For the first time in a long time, she'd lost faith in what she did. The whole point about chasing criminals was that you made a difference. With these two cases, she felt like she was getting nowhere.

'What's that then?' Kelly asked.

'The zip found with Brian's body. It matches a manufacturer of the brand of jeans Brian was wearing. His mother gave a very good description at the time and it matches the pair you saw when you visited the house.' Kate showed her a photo.

'His bedroom was like a mausoleum,' Kelly said.

'Okay, cheery, let's crack on shall we?' Kate said. 'The synthetic label has been traced to a factory in Leeds. It matches the ones sewn into T-shirts like the one Brian was wearing.'

'Kate, I know I'm being a moody bitch at the moment, and please allow me that, though I do apologise. The thing is, neither the zip nor the label gives us any leads as to who murdered Brian, they simply confirm his identity, which we've already done.'

'Ah, now, that's where you're mistaken,' Kate said.

Kelly looked at her. 'What?'

Kate grinned. 'It's a long shot, but last year a team of specialists at Cambridge University managed to extract DNA from the fur worn by a very dead caveman – or woman – five thousand years ago.'

'And what has this got to do with us?' Kelly asked.

'I've spoken to one of the scientists and he was very excited about the leather binding around Brian's wrist. Because it's organic matter, and if it was buried in the lake bed – which it might have been for many years – it could have been preserved in perfect condition. He wants to test it.'

Kelly paused. She didn't know whether to laugh or cry. The chances of such a test being successful was, in her mind, poor at best. But then science was developing new tools for analysis all the time. She'd had cases where DNA had degraded in a matter of days, on pillowcases, underpants, and on bedsheets. But Kate looked as though she really believed in the possibility that some might have survived intact on fragments of leather that had been under a lake for twenty-four years.

'Fucking hell,' Kelly said finally.

'I knew you'd be happy.'

Chapter 42

Joanne Cooper lived ten minutes away from Eden House.

Kelly willed the Merc's air-conditioning to kick in quicker. It had worked hard these past few weeks, and she was beginning to think it was losing its benefit, or was it simply getting hotter? She might have been better off walking, but now, as she waited at a set of traffic lights, gazing towards the national park, imagining which hill she'd love to hike up to find a wild pool of cold moving water, she resigned herself to the wait.

Joanne's house was a pretty end of terrace which was well kept, with an attractive front garden. It reminded her of her mother's garden. Wendy had been a keen horticulturist and planted every inch of spare space she had, inside and out. The windowsills in their tiny house were always filled with scented flowers and herbs, and pots laden with colour lined the paths to the front and back. Kelly thought there was something special about people who looked after plants so well. They were patient and kind. But then she'd known serial killers who had a penchant for growing things too.

She parked on the street and knocked on the door. Joanne was retired so there was a fair chance Kelly would catch her in. She did. The woman answered the door and held it open.

'I expected you'd want to talk to me at some point,' Joanne said.

She wore a pair of brightly coloured shorts and a vest top, and she had gloves over her hands, and she held what looked like a dead plant. Kelly peered at it.

'Oh, it's got life in it yet,' Joanne said. 'They don't like being moved, especially in this heat, but I'm trying it somewhere shady, poor thing. Come in.'

Kelly followed her inside and closed the door behind her. Joanne took off her gloves and offered Kelly a cold drink. She accepted a glass of iced tea, and she watched Joanne clear a small table out the back, which was positioned in a lovely patio area, in full sun, but protected by an umbrella. There was no breeze and Kelly gulped the cold liquid. They sat down.

'How can I help?'

'How are the funeral arrangements coming on?' Kelly asked.

'You didn't come here to ask me that. What do you really want?'

'I want to know why and when you fell out with your brother. Michelle Parkinson seems to think that you were still close.'

'Michelle Parkinson knows nothing. She's the same little girl with an abused mind and lost soul she always was. Jason fell for it. We all did.'

'Fell for it? Are you telling me that Michelle's help in my inquiries is disingenuous?'

'Yes.' Joanne smiled. 'I was done with playing games years ago. It's why – to answer your question – I fell out with my brother. It wasn't so much a falling out, it was more me telling him to grow up and him refusing. You get tired of people leaning on you after a while. It was shortly after Brian disappeared actually. He changed. I knew he'd been up to something. If he wasn't directly involved, then he knew exactly what happened to the lad.'

'Why didn't you tell us?'

'You never asked.'

'Okay, Joanne, I'm all ears. Is that why Jason turned so spectacularly to drink after he dropped out of college?'

'Bang on. Of course he did. He felt responsible. I tried to get him on the straight and narrow, so did Michelle, but he knew he'd done something so bad that he just couldn't get over it. Being drunk was so much less painful, you see, than being sober.'

'I get it,' Kelly said. 'Did he ever tell you what happened?'

'He didn't need to. I remember that night clearly. He came into my room sobbing. He was covered in mud. I yelled at him to get off my bed, he made it so dirty, I had to wash it. My mam screamed at us both. There was a time much later when he was drunk to the point of close to blackout and I put him to bed. He said he was sorry. He cried. He cried a lot. It was the burden, you see. He tried to hold down jobs, even getting married and becoming a father. None of it was strong enough to make him heal. He just couldn't get over it.'

Kelly listened, not wanting to interrupt lest Joanne stop talking, but she had so many questions.

'What made me so angry, and I regret it now, was that I didn't properly listen to what he was telling me. I always thought he'd done it on his own, you see. I figured they fought over Michelle – it's always a girl isn't it – and it was an accident. But then he told me.'

Kelly waited. She held her glass tightly.

'They were just going to warn him.'

Kelly's stomach tightened and she wished there was more air to breathe.

'I never believed Jason was capable of being directly involved.' Joanne paused.

Kelly recalled from Henry's simulation that it was an adult who struck the killing blow. She waited for Joanne to continue but worried that she might have suddenly changed her mind about sharing the burden of the secret that had tortured her about her brother for so many years.

'I don't know…' Joanne said finally.

'What did Jason do, Joanne?'

Joanne looked up and stared directly at Kelly.

'Oh, it wasn't Jason who did it. It was Michelle.'

'Michelle Parkinson?' Kelly asked, stunned.

'You thought I was going to tell you that Jason did it and then his guilt drove him into a pit of despair, didn't you? Well, that's what I believed too. Until Michelle told me something by accident. I suppose after years of lying, it catches up with you. I told him to go to the police, and finally get rid of the guilt he'd been carrying all these years, but he wouldn't have it, in fact, it made him worse, what I told him.'

'What did you tell him?'

'Well, you see, I followed the story very closely. I would have done anything to get my brother on the straight and narrow. I was young and naïve. In the end it was Michelle's own mouth that sealed it for me.'

'How?'

'I always knew that Michelle was desperately unhappy, and that makes kids do weird things, doesn't it? It wasn't Michelle who wanted her father punished. It was Brian. I asked her straight out, oh years ago now, if she ever thought of Brian and what happened to him. At this point, Jason hadn't told me anything at all. I was asking Michelle for help getting him home one night. I drove to The Swan Inn and got him in the car. She helped me. Then he started mumbling about Brian. I couldn't fathom what he was saying until Michelle said it wasn't his fault that he didn't know Brian couldn't swim.'

'I don't follow,' Kelly said.

'Brian sat out of swimming lessons at school. He never learned. Jason thought he faked it to get out of the lesson. But that teacher knew, didn't he? And so did Michelle, because they'd pushed him in on a school trip, and he had to be rescued by the teacher. Michelle was the one who said to Jason, and I'll never forget her words, "no one can swim with their hands tied behind their back". Then it struck me as odd because in the paper this week, an article mentioned that teacher telling the press that Brian couldn't swim, and it was the same one who pulled him out of the lake on the school trip. I remember his name because he was the one that Jason was always hanging around with. It was Mr Thompson. Steve they called him. Why would he say that? Unless he knew how he died?'

Kelly didn't follow her logic at all, and her initial excitement turned to disappointment as she realised that Joanne was piecing together cobbled together parts of stories and searching desperately for a reason why her brother went so spectacularly off the rails.

'Then I saw Brian's picture, and it all made sense. Jason, the night I picked him up in a state, was mumbling about a leather jacket he wanted, that Brian would never let him borrow. And that's when Michelle said not to worry because the jacket was taken care of the same as Brian, and nobody would ever wear it again. Jason actually cried. I still believed he was with it enough to listen to what she was saying, because he said she ruined it by cutting it up.'

'I still don't follow,' Kelly said, feeling inadequate that she couldn't legitimise Joanne's concerns, and stupid for missing something vital.

'The jacket! Look,' she said. Joanne handed her a newspaper from this week and showed Kelly the old photo of Brian, in his leather jacket, smiling at the camera. The photo had been taken by Donald, his father, and it made Kelly's guts churn every time she looked at it.

'That's what he was wearing wasn't it?'

'Yes,' Kelly said.

'Jason said he never stood a chance because he couldn't swim.'

'Yes.'

'And Michelle said you can't swim with your hands tied behind your back.'

'Yes.'

Kelly floundered.

'Look!' Joanne thrust the newspaper further into her face. 'I'm no expert, but why did the artist draw a skeleton with the arms behind his back?'

Kelly peered at the black and white sketch beside the article. She'd thought it was a stock picture of a skeleton, which is what newspapers usually produced when they had nothing else. But this one looked startlingly similar to how she'd first seen Brian's body. It was possible that a journalist had seen the remains before the forensic tent was erected, it happened all the time. They should adhere to an unspoken code not to print what they see, but it's tempting to break that code when column inches are sparse.

'That information wasn't released,' Kelly said weakly.

'So, they made it up? Pretty accurate then? I can tell by your face. That's how he was found, wasn't it? I remember you now. You went to school with Jason, didn't you? I knew I recognised your face. Jason talked about you. You were the one who left, made good, got some high up job in London? So you saw Brian being chucked into the lake by Mr Thompson?'

Kelly's temples throbbed. The memory that had bothered her all week wasn't the one of Michelle wet through and crying, or Brian's face after Jason hit him, or the fact that the trip had ended early. It was the image of Brian struggling to keep afloat in the cold water of Derwent, after being chucked in by the teacher in charge of everybody's life for a week. Her initial recollection had him falling in, but now she saw it clearly. She felt nauseous. She'd apparently blocked it out. She'd been so focused on the detail that she'd questioned her memory of the events of that year; they were hidden behind the fog of desperately wanting to fit in, of admiring the main players from afar, of ignoring the signals and assuming that those in authority did what they were supposed to do: tell the truth.

When she should have trusted her own instinct and recollections, they'd let her down. Now Joanne had reminded her she saw it in full, played out for everybody to see. Brian, and the panic setting in as he flailed about. Mr Thompson laughing. All the way to the present day and what Joanne had just told her.

She swallowed and finished her drink.

'How well do you know Michelle Parkinson? She gave me the impression that you two were friends,' Kelly asked.

'Well, a bit like the devil himself, I keep my friends close but my enemies closer. If you're asking me if Michelle is capable of killing Jason to finally shut him up, then the answer is absolutely yes. But the person you need to find is that teacher. The night I picked Jason and Michelle up from The Swan Inn, that's who he was with, and he came out to the pub car park and watched as we loaded Jason in the car. I asked Michelle who he was and she said that I should forget I ever saw him. I told her I'm not scared of anyone.'

'What did she say?'

'She said you should be.'

Chapter 43

Carleton Hall was the type of building that looked as though it should have housed some top secret government department in the 1960s in its day. It had a grand sandstone façade, onto which had been added huge blocks of concrete extensions, and the windows were so high that one might be forgiven for thinking it had in fact been converted into a mental asylum. Sometimes that's exactly how Kelly felt when she visited, to see Superintendent Andrew Harris.

She didn't make it a regular occurrence, but she felt, and Kate agreed, that this was one of those times she needed to see him face to face.

Two high-profile murder cases were one thing, but the sniff of police incompetence that was swiftly turning into a tornado cloud of screw-up after screw-up, documented and corroborated by a growing number of witnesses, was quite another.

Andrew Harris was a tall, thickset man, who'd be perfect for a TV show on brawny, no-nonsense coppers. He had a deep voice but kind eyes and he smiled broadly when Kelly was shown into his office. She was offered tea or coffee, but declined both. She wanted to get back to Eden House in time to face Steve Thompson.

'Sir, I've got a problem,' she said. She sat in front of him; the desk separated them. He leant forward on his elbows.

'Of course you have, otherwise you wouldn't be here. Is this about your two current cases? The boy found in Thirlmere and the murder of an old school pal of yours?'

'I knew the one found in Thirlmere too,' she said.

'Too close? I can move you onto something else, the drug squad is crying out for help in Barrow-in-Furness at the moment, if that's what you'd like. Kate Umshaw can step up, and I hear Dan Houghton is a good operator.'

'No, sir. I don't want out. My concerns are personal, but it's not because I can't cope. You may or may not be aware that my stepfather worked this patch many years ago. John Porter. Good reputation. Retired a bit of a hero.'

'I recognise the name, obviously. I hadn't made the connection.'

She gave him a brief outline of her cases, as well as how one of them had been handled twenty-four years ago. She was speaking for twenty minutes and after she'd finished, she felt better for having got it all off her chest.

Harris sat back in his chair. 'Crikey,' he said.

'If the press get hold of it, the shit will hit the fan, sir.'

'Indeed. What have you got?'

'I've certainly got some tricky witnesses. I have no idea who is telling the truth, if any of them. But my problem is physical evidence.'

She told him about what Kate had said about a new method of extracting DNA from ancient organic material.

'1997 isn't exactly ancient,' he said.

'That's not my point, sir. It's surprising but not impossible that DNA wouldn't completely degrade in that underwater environment. I'd like to give it a go. I think the only way I'm going to get to the bottom of these cases, which I now think are connected by Jason Cooper, is through the physical evidence. It'll be costly, and lengthy, but worth it.'

'Do it.'

'What about the original investigation, sir?'

'How worried are you that it was purposely evasive and not just sloppy?'

'It doesn't really matter what I think, sir. It's historic abuse of process, but there's no way I could prove that at a tribunal, or through public inquiry.'

'Leave it with me. I want a report to me by the end of play today with what you've got. I'll make a decision then. Did you not see eye to eye with your stepdad then?' he asked.

'On the contrary, I thought he was a legend, professionally. It was only recently I found out he wasn't my real father. There are a few of the old guard left who he used to speak of. I know a couple of them still live in Cumbria – may I put out a few feelers, sir?'

He exhaled. 'Are you prepared for this particular can of worms, Porter?'

He only called her Porter when he was terribly serious.

'If we hand it over to internal affairs without any evidence, they'll reject it as a complete waste of their time. The least I can do is do a bit of digging, it could help with my current cases anyway.'

'Okay, agreed. How close are we to wrapping up these murders?'

Kelly sighed. 'The 1997 case, not really, sir. We have such little physical evidence, it's almost impossible to think that we'll get a suspect anytime soon, without a confession. Besides, our main suspect may already be dead. It's Jason Cooper.'

'And his death?'

'Somebody who was involved in Brian's death too perhaps, cleaning up after the body was found. It's more likely we get resolution on that case because at least we have his body and a crime scene.'

He nodded. 'Apart from all that, how is your team? We need to discuss appraisal before the year end.'

'Yes, sir, I'll schedule a meeting. If that's all, I'll get going,' she said, standing up to leave.

As she walked back to her car, she called Kate and instructed her to go ahead with the Cambridge University lab technicians, with Andrew's blessing. It was their only hope of finding enough DNA to produce any kind of profile to feed into the databases. The slivers of leather which had held Brian's arms together for twenty-four years could turn out to be vital, or a complete waste of time, but she had to try. Maybe they'd get lucky and the cold water and silty bottom of Brian's final resting place would have preserved some organic matter. That would be miracle one. Then all they needed was for whoever tied him up to be stupid enough not to wear gloves. That would be miracle number two.

Chapter 44

The team that was tasked with visiting a widowed gentleman who collected old model Fords as a hobby consisted of two uniformed officers and two forensic crime scene investigators. He was expecting them. He lived on a sprawling patch of land on the edge of a forest ten miles out of Aberdeen, and he explained to them, in fine detail, how he'd chosen it after his wife died, precisely because he could build a garage big enough to fit all his cars.

He showed them to his garages outside, which were spread over a vast area that looked as though it could have once been stables. He took them to the outbuildings and clicked a key fob, opening one of the large electric doors.

'There she is. A 1980s retro masterpiece,' he said, beaming with pride. 'I'll leave the key with you,' he added, opening the front doors and the boot compartment. 'Go easy with her.'

The officers went into the garage and walked around the vehicle. Despite its age and model, it could have just rolled off the production line; however, the boxy lines and tiny period features gave away its vintage. It looked pristine, and one of the forensic officers frowned.

'Can you tell us exactly what you've had done to it?'

'It's been recalibrated, in so much as the engine was taken out and cleaned and updated. Some of the fixtures were replaced, the front seats were in terrible shape. The steering wheel has been replaced, I found an exact match on eBay.' He beamed. 'Of course, it's been valeted, and the clutch went, oh, and I managed to find a vintage radio that worked.'

'We'll get to work, thank you,' one of the forensic officers said to the man.

'Can I get you some tea?' he offered. 'It's thirsty work in those suits, I bet,' he added.

The officers accepted, and he disappeared back inside the house.

The two forensic officers made cursory examinations of the vehicle as the uniformed officers kept guard, not that they were expecting anyone to interrupt their work.

'It looks as though it's been through a sterile sheep dip.'

'Aye.'

'Ach, well, let's have a look.'

They'd brought equipment capable of detecting stains under certain light conditions, as well as evidence bags, various sizes of tweezers, different types of magnification kits, and boxes to take away anything of consequence, though looking at the vehicle, they doubted they'd find much. They'd been briefed that the information was to be used in a case from 1997. But they'd give it their best shot. One walked to the rear and peered inside the boot. He set up lights that could detect organic matter once luminol spray was cast over the suspected area. The water-based solution was capable of detecting blood even when it had been diluted thousands of times. Now a known carcinogen, the officers needed to use the spray carefully, and they wore masks. As one worked inside the boot, the other concentrated on the back seats. Any processing of the front of the vehicle would be useless, thanks to the recent refurbishment. First, they had to make the scene dark, and so one of the uniformed officers was tasked with asking the owner to close the garage door, then they planned to erect a tent made from blackout material. It was meticulous work and took time, but one didn't go into the field of forensics expecting instant gratification.

Once they'd set up the site and finished their tea, they were ready to begin work. The interior of the car was now acceptably dark and the first tiny dots and stains appeared before the scient-ists. They weren't in the business of becoming overly emotional

234

or excitable when the little blue marks appeared, because they knew from experience that other organic material, as well as chemicals, reacted with the spray. It was their job to focus on the areas picked up as potential bloodstains and cut pieces of material from the specimen and prepare them to be sent to the lab.

After they were satisfied that the interior of the car had been combed for traces of blood proteins, they prepared for their next assessment, which was to search, by hand, for any foreign articles adhered to the soft furnishings and sharp edges of the car's interior. Even the best valeting services couldn't get rid of tiny microscopic fibres left behind by previous owners, and they hoped at least to discover something that might have been inside the car for a long time. Today's cars were designed to be aesthetically pleasing as well as to work more efficiently; however, four decades ago, things were more rustic, especially on a standard model such as this one. They weren't talking Bentley here, this was a run-of-the-mill basic model Ford, and under the furnishings they found all manner of protruding promontories on which a whole array of human evidence could become snagged.

Occasionally, every hour or so, the forensic officers stretched their backs and took a break to wipe their brows. Working in full overalls, as well as head cover and masks, was sweaty and unpleasant, but the job, ideally, must be performed in one shift, otherwise they'd have to sterilise equipment, check for cross-contamination and start from scratch.

They crawled into tight spaces, removed carpets and chairs, and unscrewed nuts and bolts. A quick brush and clean with washing solutions, performed by kids on minimum wage, was never going to get rid of detritus in such confined spaces, and they remained hopeful that they'd find something of worth to pass on. They knew their work was for a criminal case, and rarely did they see their efforts out to their final conclusion, unless they were invited into a courtroom, but their reward was in doing their job properly.

It was late afternoon when one of them called out to the other, to point out a find that at first glance looked promising.

Snagged on a miniscule shard of metal, protruding perpendicular to a bolt underneath the front passenger seat, was a twist of fibres which looked synthetic. It could have been a manufacturing oversight, or simply that the packaging was taken off in haste; it could even be from a cleaning cloth, or a footwell carpet taken out to clean. It was obvious that the fixture was original and hadn't been replaced with the seats. But whatever it was, it was worth looking at under a microscope in the lab, and the fibres were carefully extracted and placed into a plastic bag, sealed, labelled and collected.

By the time they'd finished, they stood back, and smiled, taking off drenched head coverings and gloves. Their faces were pink and the uniformed officers stared at them, yawning.

The owner was allowed back into the garage when they were done, and he was reassured that the company would reimburse him for the damage caused to the soft furnishings, via the Cumbria Constabulary. However, it quickly became apparent that he hadn't understood the content of the warrant, and he was difficult to console. The officers, who'd hoped to finish their work and get home, had to stay behind for another hour or so to calm the man down and offer him guarantees that his beloved time-piece hadn't been utterly ruined.

'It's cosmetic,' one said to him. 'That's the whole point. It's not structural, you should see some of the vehicles we've worked on,' he added with a smile.

It didn't help.

Chapter 45

Friday nights had never been a huge deal to Kelly. While other people in her peer group had prepared excitedly for the night ahead, visiting the gym last minute, buying a new outfit, and gathering with friends early to drink and gossip, she'd always been finalising cases, visiting witnesses or facing a long weekend trawling over a new case. It had never bothered her.

What did bother her was that Steve Thompson hadn't turned up to his interview, and he wasn't answering his phone. Westmorland Comprehensive said he was off sick.

She sat on a curious police precipice that faced all investigating officers at some point during every serious case: when does a person of interest become a suspect, and thus require an arrest warrant? Steve Thompson hadn't yet crossed that threshold and so to go charging about, using valuable resources, trying to find him on a Friday night, wasn't the best use of her time. They urgently needed to speak to him, and that's what was being put out by the press office, along with his photo; however, there were different grades of urgency for any police force. Speaking to Steve Thompson was quickly becoming a more serious priority for them, but Kelly also knew that collecting data and evidence was her number one concern.

After all, what use is a suspect without any pressure to bring down on them? Zero. Shooting from the hip and dragging suspects in too early only resulted in them being released without charge. They needed something solid.

More pressing was the report staring at her from her computer screen. It had been sent by the Lancaster University team who'd been responsible for excavating the site where Brian had been found. She remembered Henry saying they should take at least a foot of soil from around the body, and that's why Brian had been placed inside a wooden coffin. Now, the pedologist's report had come back and Kelly couldn't quite believe what she was reading.

She dealt with scientists all the time, but their jargon often confused her and she had to call them for clarification. On first reading, the report seemed to suggest that Brian had in fact been buried in a shallow grave, definitely not underwater: after all, who could do that?

Something the woman who lived near Thirlmere had said jarred her. She said the water company had been doing maintenance, back on that July night in 1997. She googled the date and 'Thirlmere reservoir' and a few articles popped up, but the top one was of most interest.

'Reservoir drained for hundred-year maintenance: locals livid.'

She read the whole piece, stopping to check a few facts, and found that in 1997, some 120 years after the Manchester Corporation had gained permission from Parliament to extend and dam the original Wyburn Lake to provide extra water for Britain's growing cities, routine maintenance, under the new ownership of United Utilities, had commenced that spring. It involved cleaning and checking the boundaries and equipment that had lain undisturbed under the huge body of water since the nineteenth century. It wasn't that uncommon, Kelly learnt. Attached to the article were links to photos and she clicked on them, examining them one by one, piecing together a picture of the enormous task and what it involved. It cost the utility company millions of pounds but compared to their profits every year it was literally a drop in the ocean, or lake, at least.

She expanded one photo in particular. It was useful in that it showed the extent of the draining. Obviously, a reservoir of over three square kilometres in area couldn't be completely drained, and it didn't have to be, to reveal the main structures which needed to be serviced and modernised. She'd read in the article that even after a hundred or so years, the engineering was in good shape and the expense was worth it to establish its structure. But what she saw now, when she zoomed in, was that the place where Brian's body had been found had been exposed during the work. It was a dry bank, close to the remaining waterline, for sure, but exposed nonetheless.

She referred back to the findings of the pedologist and the part where she specifically stated that Brian's body was encased in soil that was not consistent with the surrounding lake bed. In other words, it had been disturbed. There were various factors that indicated if a soil type matched up with its surrounding deposits, such as chemical traces, animal matter, microorganisms, colour and density, and in Brian's case, the soil around his body didn't match the rest of the area where he'd been excavated. The differences were minute, but there. In other words, it was entirely possible that Brian had been buried in a shallow grave, while the water level was low during maintenance work, and covered over, even with the same soil, but digging it and compacting it had changed its structure and composition forever.

'Holy shit,' she said under her breath.

It was the most scientific evidence they had so far on how Brian's body stayed at the bottom of the lake bed for all these years without ever resurfacing during decomposition, and thus holding together, even with the restraints that bound his hands remaining behind his back. Whoever had dug the grave must also have been familiar with the bed of the reservoir, and where it was manmade, and where it was a decent depth of soil.

She called Henry.

She soon got the impression that her father and the doctor were in the pub, discussing their findings on her cases. He asked her to join them.

'As much as I'd love to, I'm a bit bogged down,' she said. The pun was unintentional. She told Henry about the pedologist's report.

'Well, now that's fascinating, and yes, it would explain why the body didn't move from the bottom.' She heard him relay the findings to Ted, and waited patiently. They also discussed Brian's lack of clothing and agreed that he must have been stripped.

'If he'd been clothed and buried, even below the water table, his clothes would still be on him, intact, more or less. We were just discussing the results of the scan.'

'Scan?'

'I ordered a 3D scan of the earth inside our temporary coffin when it first came in to the mortuary. You wouldn't believe the perfect picture it threw out to us. I have a copy here with me. It's exactly as you say. The earth around the remains is compacted like a grave site. I think you have your answer. And if the lake bed was exposed, even if it was still soaked from the reservoir, then it would have been rather easy to dig, wouldn't it?'

Kelly sat back in her chair.

Brian had been buried. And the grave more than likely dug with the spade that killed him. She fancied betting rather a lot of money that a local geography teacher was likely to have known about the draining of a huge reservoir, and, understandably, would have taken a keen interest in the proceedings.

With the afternoon opening up before her, in the absence of being able to speak to Steve Thompson, Kelly had tasked Dan, after he'd finished with the witness at Thirlmere, to pay another visit to Dave Crawley, to challenge him on what Paul had told her last night, that Dave had criminal dealings with the teacher before being caught and sent to prison. She was keen to hear his feedback, and as she sat pondering the new information she'd been given, he poked his head around her door.

'I'm back. He's a nasty piece of work, isn't he?' he said.

'Sure is.'

'You went to school with some rough 'uns. I'm from Glasgow, I thought I'd seen it all.'

'It was a vintage year,' Kelly said. 'Sit down. Tell me about our curtain twitcher first,' she said.

'She took me across the road to the lake and pointed out where the work was carried out and walked me through where the car turned. She'd make a cracking witness.'

'Did you show her the photos of Thompson, Brian and Jason?'

'Yep, she couldn't say for sure but she does remember blonde hair, and she said an adult male was definitely driving, not a kid.'

'Okay, what about the lovely Dave? Did you get a list of his incoming and outgoing post?'

'Sure did. He's a popular man. He gets parcels every Friday for his canteen, and the screw who I talked to said he gets the most sent to him by a mile, and of course he trades half of it inside.'

'Any names?'

'Oh yes. A regular parcel from Steve Thompson.'

'He doesn't even disguise his name?'

'Nope. Cock swinging in the breeze, boss, as confident as you like. He rather enjoyed it when I pointed it out. He was cagey about everything else, obviously. And he asked loads of questions about what we knew, rather than answer too many.'

'Typical Dave, trying to sniff out where he stands in all of this,' Kelly said. 'What's usually in the parcels?'

'Gucci clobber. All the stuff that sells well inside, like shampoo, sweets, vapes, toilet roll, porn. All the regular booty. The scanners pick up the odd contraband, though as we know from our previous visits to Highton, there's more than one way to skin a cat, and these boys are desperate inside aren't they?'

'So he thinks he's in charge. Classic Dave,' Kelly said.

'Aye. I noticed that. He's the type to tell a story, isn't he?' he said. 'Told me all sorts of terrible things that his associates had done, but not him, of course. He said that Steve Thompson had given him contacts who he found useful.'

'Such as?'

Kelly dreaded what Dan was about to tell her, as the period in question covered her engagement to Dave Crawley.

'Like all cons, he has loads of time to construct elaborate stories inside,' Dan said. 'It doesn't look good for our Mr Thompson, though Crawley could be toying with us, he loves the attention. So, the only thing I've been able to verify is that Thompson's family owned a place near Cockermouth, tucked away, and that it passed to Thompson when they died. His father died back in 1993 and his mother two years later, then the property was rented by Crawley & Son from 2005 onwards.'

'Wait, this didn't come out in the trial against Crawley,' Kelly said.

'Probably well hidden, and I'm guessing it wasn't a legitimate transaction,' Dan said.

'And Paul told me they were terrified he'd be ruled a hostile witness, no wonder. We can only speculate what the property was used for,' Kelly said, knowing that whatever the Crawley's business interest in an empty isolated property, it wouldn't be benign. 'Anything else?'

'He revelled in talking about John Porter.'

Dan looked sheepish and Kelly felt her cheeks go slightly pink. It wasn't embarrassment, more dread that the revelations about the man who'd raised her were not over yet. Andrew Harris had given her permission to look into the accusations, but they were piling up and she felt like handing it over to an independent third party.

'Like what?' she asked Dan.

'He said that Thompson's father was a copper – a bent one, and a boozer – and he trained with John Porter back in the late Sixties, in Preston, and they remained close.'

Kelly felt as though she'd been smacked in the face. Dan carried on.

'I've looked into it and he's telling the truth. Bobbie Thompson was honourably discharged in 1989.'

'In 1989, that could have meant got rid of. Did he work here?'

'No, he worked in Lancashire.'

'But perhaps they remained close, is that what Dave was getting at? Bad apples and all that?'

'I reckon so. But this is where he got very smug, boss. He told me all this so he could get to his crescendo.'

Kelly braced herself. 'Which was?'

'Well, it got me thinking about how Thompson has no previous on our system, and doesn't appear to have been flagged up before for criminal links. Now, I thought, okay, that's fairly normal, some people are super clever at evading detection, but it still doesn't fit with the profile we're exploring for him. You'd think that a dodgy teacher would have been weeded out of the profession, given the rounds of disclosures they have to do now. Either that or he's extremely lucky.'

'Or he's innocent,' Kelly said, thinking about how much she looked up to her geography teacher and how noble he was in her memory, the memory that had let her down lately. Wasn't that the unique talent of manipulators? They knew how to play people and were always one step ahead.

'We know that regular police checks didn't even exist for teachers before 2002, unless they were convicted child abusers on List 99, so either he was excellent at diverting attention away from himself, or like you say, it's all bollocks, or...'

'Or?'

'He had someone vouching for him.'

The penny dropped. Kelly put her elbows on her desk and covered her face with her hands. She peeked over her fingers at Dan, waiting.

'Like a senior copper?'

'In the borough he was applying to teach in,' Dan added.

Kelly's gut sank to her toes. Now it made sense.

They knew that Thompson had worked in the same school for over thirty years, never moving on, and keeping below the radar, even after the disclosure and barring service replaced the criminal records bureau in 2012. Thompson didn't need to have committed a crime, that wasn't the point. But it made him untouchable and this, Kelly knew, was what Michelle, Paul and Dave all alluded to.

If Brian wanted to expose him...

'Mrs Gooch has been headmistress there for years. Why are schools so full of dinosaurs?' Kelly asked weakly.

'Might do us a favour,' Dan said.

'He's got to have some kind of discipline record, he can't have remained squeaky clean all these years, when even the kids knew he was dodgy. I'll talk to her,' Kelly said.

Dan stood up.

'Dan?' she said. 'Crawley has been sitting on this for years. Did you get the impression he wants to tell us in instalments?'

'Definitely. And he's probably not done yet. He's loving every minute, especially the involvement of John Porter.'

When Dan had left her office and closed the door, Kelly sat for a moment, remembering the close relationship her fiancé had with her father, all those years ago. It made her feel as though she wanted to either vomit, or cause serious harm to anything she could get her hands on.

Chapter 46

Westmorland Comprehensive School had finished its work for the week, and there were few cars left in the car park when Kelly arrived. Mrs Gooch was expecting her. For the second time this week, she felt like a sixteen-year-old who was in trouble.

The school was a mixture of grand old grammar school wealth and ugly prefabricated additions. It was built over three storeys, and sat in impressive surroundings, with plenty of fields and space for children to run around and expend their energy. On the outside, it looked like a fine institution that ticked all the boxes. Any parent would feel comfortable with the façade of propriety and sturdiness. But it was a bit like the people who Kelly's mother told her wore a fur coat and no knickers. It impressed on the outside but was lacking on the inside: rotten underneath a layer of capital success.

She walked to the entrance with a mixture of trepidation and depression. One by one, her childhood memories had been smashed, or were in the process of being obliterated. Things she took for granted no longer existed, and people she'd assumed were one way, turned out to be the opposite.

She had to show ID to get through the Fort Knox-style security at the entrance and Kelly shivered at the possibility that it didn't keep out all mal-intent. Once you were through the door, you were free to roam. She felt judgemental and admonished herself because that wouldn't help her inquiries. The only way she'd get justice for Brian Miller and Jason Cooper was to dig up the truth, and that was proving to be more about the people involved than a good old-fashioned trail of evidence. Physical evidence couldn't lie, but people always did.

Mrs Gooch looked the same as she did twenty-four years ago but a little smaller, due to her age-induced kyphosis, and softer around the edges, thanks to her change in fashion. Her eyes were the same, and so was her voice, and Kelly was transported back to her assemblies, when she'd lecture the children on morals, stranger danger, and boys with fast cars. She was taken into her office and Kelly realised she'd never been there before. As a child with a good academic record and no discipline issues, she had no cause to be invited there. It was larger than she anticipated, and covered in framed photos of sporting achievements and ex-heads. It was an adult place in a children's world and Kelly saw her as a peer rather than an all-powerful master of authority.

'Congratulations, Kelly. You've done well, as we always expected.'

Her praise irritated Kelly. It alluded to the unwritten rule that good children turned out well, and that the bad kids went off the rails, as if it were scribed in some great tome hidden behind the trophies and accolades decorating the walls.

A wave of cynicism washed over her.

'I'm here to discuss Mr Thompson,' Kelly said. 'We can't find him.'

The knowledge, acquired by Dan from Dave Crawley, that one of the teacher's properties had been rented to the Crawley family – convicted people traffickers – had solved her dilemma over whether to move on him. All units in Cumbria had been alerted to intercept Steve Thompson, when and if they found him. Kelly reckoned he hadn't just gone for a hike, or taken a breather.

'He called in sick today. He's retiring after the end of this term; he's had a long and successful career.'

The platitude was bland and Kelly didn't buy it.

'What was his discipline record like?'

'What do you mean?'

'I'm interested in his track record with the children he taught. Was he heavy-handed? Did you receive any complaints about his behaviour towards them?'

Mrs Gooch fiddled with papers on her desk, flustered.

'There are some historic issues that have come to light and I was hoping you'd be able to assist us. For example, his application to work here, who was his reference?' Kelly dived straight in, expecting the worst.

'Goodness that was a long time ago.'

'But you were head, right? So you'd know. I'm looking for transparency, I don't really want to have to request a warrant on a Friday evening, if you understand me.'

'Yes, of course. We're here to cooperate fully, but the Data Protection Act—'

'Isn't relevant when the police suspect a crime had been committed.'

'A crime?' Mrs Gooch went pale.

'I'm not at liberty to discuss the details with you, I'd just like to see his employment file.'

'Right.'

The game of brinkmanship was over before it began and Mrs Gooch tapped some keys on her computer.

'Here we go. Our systems were updated three years ago, we're all online now.'

Kelly figured that Mrs Gooch would like some praise for this effort, but it wasn't forthcoming. Kelly was tired and she longed to get home and at least enjoy a bit of the weekend with her family.

Mrs Gooch got up from her chair and beckoned Kelly to take her place. It felt odd, being allowed to sit there, on the throne of power that held children's fortunes in its grasp. It was as if they were filed into two groups as they walked through the gates at the age of eleven: the losers were ushered to one side, while the well behaved, those who caused no trouble, were given a free pass to greener pastures, and allowed to go on to live lives on the right side of the law. It was a pity that the same scandalous arbitration didn't apply to adults.

She scanned Steve Thompson's file, and her gut sank to her toes.

'That's quite a list of disciplinary warnings. Why did he stop leading school trips in 1998?' she asked.

'He went into a girls' dormitory,' Mrs Gooch said quietly.

'And none of these were sackable offences?' Kelly asked.

Mrs Gooch shook her head. 'We need a file as thick as the bible to take action against a member of staff.'

'And this one wasn't thick enough?'

'We followed all school policies to the letter and they're audited and checked by the local authority every three years.'

In other words, Kelly realised, they'd done everything correctly and were thus exonerated from culpability. The establishment protecting their own. She read the list again. Thompson had been hauled in a total of seven times over misconduct issues, ranging from a parent complaining of inappropriate language in class to a complaint in 1996 from Donald and Maureen Miller registering their disgust over how the incident with Jason Cooper had been handled. Then she brought up Thompson's employment application and saw what she'd been dreading since her conversation with Dan this afternoon. Thompson's reference on his application was John Porter. Furthermore, since his hiring by the school in 1992, Thompson hadn't submitted any evidence of DBS clearance.

Because he hadn't had to.

Chapter 47

When Kelly returned to Eden House, she couldn't hide her ill temper. Her face was thunderous.

She walked to the front of the incident room, and stood in front of the whiteboard. The room fell silent.

'We've had some movement on forensics,' she said. Stick to the facts...

She brought up slides showing the maintenance work done on the reservoir back in 1997, indicating the location of Brian's body, marked with a circle. There was audible discomfort in the room. Every time she discovered new information about the case, she was finding it increasingly difficult to support the original investigation. Why did nobody notice the disturbed ground?

'Now we can say with conviction that Brian was buried, after he was stripped of his clothes, restrained and beaten to death, not necessarily in that order. Unfortunately, we still don't have a lead on where Steve Thompson is. Moving on to Jason's case, we have no trace on the locket yet.'

She brought up a photo of the beautiful piece of jewellery on the whiteboard.

'Somebody will be missing this. DC Hide has had feedback from a silver dealer in Birmingham who thinks – if it's an original, and from the photos we sent him, he thinks it is – that it was part of a significant limited production of 1910. There were only twenty made. He's estimated its worth at least a hundred thousand pounds.'

A few whistles reverberated around the room.

'Exactly. We now have an inventory from Connor's showing missing tools, and their barcodes, and we have some matches with what we found in Jason Cooper's storage shed at Parkie's. Forensics have lifted prints from the locket too. We have a negative result from the lake search, and it was thorough. Our biggest headache is motive, but if we take our persons of interest, it's looking likely that Brian Miller was intending to expose Thompson as a criminal and Jason Cooper stood to lose financially from such a move. I've applied for a warrant for a property near Cockermouth, which was left to Thompson by his parents. It's off the B5289, close to Brackenthwaite, and a squad car sent there has reported that no one is home. It could be empty. Until the warrant comes through, we can't go in, but seeing as none of Thompson's colleagues know where he is, and he's not answering his phone, it seems likely that he'd go somewhere familiar.'

'Are we treating these as one case then, boss?'

It was a uniformed officer who'd asked the question. It was a valid one, which she'd been avoiding, she realised now. She nodded.

'We've got the copycat injuries, the fact that Jason Cooper was linked to Steve Thompson through the black market – in exactly what capacity we may never know – and the witness at Thirlmere identifying the red Ford which belonged to Jason Cooper. My gut feeling is that a judge won't be happy with the circumstantial case against a successful teacher being enough at this point, and I'm hoping that with your hard work, we'll get some interesting lab results back over the weekend if we're lucky. If not, then by Monday. I've been told this afternoon that there were three different sets of prints lifted from the steering wheel of Jason's current car, the one he drove home from work on Tuesday.'

She wished she had something more to give them after all their hard work, something that could buoy the team, but all she had was a growing list of possible theories that weighed her down like mud at the bottom of a lake. Towards the end of the week, her team had grown to more than twenty officers and between them, they'd spent hundreds of hours on these cases already, and they were just getting started. It was tempting to stay in front of a computer screen all weekend, like she did when she worked for the Met, crunching data and evidence until HOLMES spurted out enough for an arrest. But she also knew that they needed to go home to their families.

It was Brian Miller's funeral tomorrow and Kelly had promised Donald and Maureen she'd be there. Henry had finished with him and the organisation had been quick. She guessed they didn't want to prolong the pain any longer. The local funeral director had made an exception for the family and hurried the arrangements through. Maureen and Donald also didn't want a huge amount of time to lapse between Brian's body being released and the date of the funeral being known about, because it would attract too much press. This way, a good chunk of that profession would be caught unawares. She knew that, so far, the arrangements had been kept to a small circle of people. She expected plenty of cameras there, but that could play to her advantage. She knew that occasionally killers liked to visit funerals to wallow in their power, and sometimes it was caught on camera. The more twisted variety just couldn't stay away from the scene, the family, and the press circus. It was a possibility that she could see something that she hadn't seen before, or that something might be picked up by a TV camera when studied later. That's how Ian Huntley had been caught, after all. After that case, the government had tightened up the vetting process, because like it or not, anything that gave such simple access to children was a magnet to people like Huntley. Finally, after 2002, anyone working with kids had to be vetted.

But not Steve Thompson.

Chapter 48

Occasionally Kelly walked into her home and felt empty. When he was there to see it, Johnny spotted it straight away: her sunken shoulders, her distant stare and her impatience with her daughter. Lizzie crashed into her legs and Kelly sighed. Before she could take the bag off her shoulder, Johnny had swept their daughter out of the walker and taken the bag, as well as given her a kiss. He glanced at her and she smiled weakly. She walked straight through to the terrace and heard Lizzie squeal, regretting her remoteness and turning back. She took Lizzie and snuggled into her neck.

There was something so pure about the smell of a child.

The wave of emotion caught her by surprise and she hid her face in her daughter's shoulder. Lizzie pointed at something and Kelly saw that it was a pile of soft toys in the middle of the room. Kelly took a deep breath and sat down with her daughter on the floor. Lizzie crawled about and passed her one toy after another. Kelly named them correctly.

'Wine?' Johnny asked.

She shook her head. 'No thanks.'

'Bath?'

'No, it's too hot.'

'Walk?'

She smiled up at him.

'Come on, you,' he said to Lizzie. 'Let's take mummy out.'

Kelly's throat constricted but she daren't let go in front of her young child. She wanted to collapse in a heap and let the tears come, in great sobs, because the man she'd grown up thinking was her father had let her down. The sense of abandonment was all consuming. As a woman in her forties now, she understood the gulf between what children see, and the truth. They look up to their adults in awe and trust, never knowing what is hidden behind the smiles.

Now she knew.

Johnny plonked Lizzie back in her walker as he gathered the kind of stuff parents need for a hike with an eleven-month-old: layers, sun cream, hat, snacks, nappies, and the like. Kelly dragged herself up off the floor and went upstairs, turning at the banister.

'Where's Dad?' she asked. The word itself was innocent, but the connotations were overwhelming and she was aware of hot tears gathering at the corners of her eyes.

Dad.

She'd known two. One was aloof, important, terrifyingly successful and proud. The other was soft, kind, loving and endlessly supportive.

'He's taken Henry to The Crown. Fancy a long one with a pub at the end?' Johnny asked.

She nodded and went upstairs, closing her bedroom door. Then the tears came. She scrunched her face up into her pillow and let out her soul into the muffled softness of the thick material. Her shoulders heaved and she screamed silently into the cover. Her head rang with the disappointment of loyalty misplaced. It wasn't that John Porter being less than the legend he was lauded as had changed his position in her heart, as the man who'd guided her and taken care of her and encouraged her, it was something else. More that this was the moment that she realised that people, including one's own family, are never as they seem. It was a flash of clarity that no one could ever attain perfection. The fact that they were all flawed struck her. But as she sat there, the ebb and flow of hormones regulating her temperament was turning into a soft swell rather than a tsunami. She felt her body calm and the tears stopped. She put down the pillow and walked to the bathroom and wiped her eyes, looking in the mirror to make sure she didn't have mascara all over her face. Johnny knew her well enough to work out that she'd been crying, but she also knew that he wouldn't comment.

She changed quickly, suddenly galvanised by the fact that they were about to get out in the fresh air. At least for the next few hours, depending on where they went, she would be at liberty to forget any worries or sense of dread that had built up during the last week. There was nothing she could do about any of it. She had one job. That was to solve a puzzle and seek the truth. It was all she could think of as she pulled a vest over her head. She wore shorts because the evening was beautifully warm still.

When she'd been in her twenties, there was no way she'd leave an investigation on a Friday night. She'd have gone to the pub with her colleagues in the murder squad, and drunk her anxieties away, only for them to return tenfold in the morning, as well as probably waking up with a copper in her bed. Those days were well and truly gone and she didn't mourn them. Her toolbox had been different then, that was all.

Now, she felt like a grown-up, getting ready to connect with nature and take time for self-care, like all the books said. It made her smile and she felt perkier. The weight of the two murder investigations lifted a little and she felt lighter in step. When she returned downstairs, Johnny had Lizzie ready and he'd pulled on some walking boots. He passed her sunglasses and she filled a couple of water bottles.

'Where do you fancy?' he asked.

'You decide,' she said.

'I've put swimming stuff in just in case it's warm later.'

She smiled at him and they checked they had everything they needed.

'Wallet, keys, nappy, cream...'

'Fighting order,' Johnny said, and winked.

Satisfied they could survive the next few hours with an eleven-month-old and not have to find a shop, they left and got into Johnny's jeep. Kelly fastened Lizzie into her car seat and the girl kicked her feet in excitement. Kelly kissed her. They weren't the sort of parents who let strict bedtimes rule their lives, and the evening was beautiful.

Johnny drove out of Pooley Bridge and along Ullswater's south shore while Kelly stared out of the window at the tourists ambling along the lakeside. The route was notorious for being blocked in summer because of its isolation, and also because of the plethora of campsites along there, towards Howtown. It was served only by this road and the steamer coming to and from Glenridding.

The sun was descending in the bright blue sky and she saw joy on the faces of people who'd been walking in the area all day long, looking to get back to their accommodation for the evening. Tomorrow was changeover day and the thousands of holiday lets in the area would be rapidly cleaned and made ready for a new crowd of visitors descending on the Lakes. Johnny waited for a large truck to pass and followed the road all the way to St Martin's Church, which was booked all through the summer with weddings, dreamt of by romantics. He carried on a little further and parked in the car park which sat at the foot of the ascent up to Place Fell, close to Dale Beck. Johnny knew it was one of her favourites, especially from this side. One could easily hike up it from Patterdale on the west side, but from Martindale, it was longer and prettier because you could see the lake all the way up. Lizzie seemed to share her excitement because she kicked ferociously when they stopped the car. She threw the toy she was holding and it landed at Kelly's feet. She picked it up and breathed out, releasing her stress.

The evening was still warm when they began their ascent. The route was steep from the very beginning, which was one of the things Kelly loved about the mountains in the Lake District. There was no trekking across tundra to the foot of peaks; it was all there in front of you from the first minute, the first breath. Heavy exhalation replaced conversation and was a therapy all its own.

Halfway up, Johnny began to talk to her.

'Whatever it is, you got it,' he said.

It was his opening gambit. She smiled as she felt her thighs burning. Johnny carried Lizzie in the backpack and she led the way. Being in front meant that he couldn't see her face. His platitudes could have been taken as shallow by anybody who didn't know him. He might be taken for a typical bloke trying to cheer up his missus. But she knew that this was just the beginning, and that he'd been thinking about her all afternoon. The knowledge softened her edges. She placed her hands on her thighs for support with every steepening step. The strength of her body flooded her system with pheromones and she marvelled at how easy it was to wash away the worries of a shitty week just by breathing in fresh air at the top of a mountain.

'I never knew John,' he said. 'But from what I do know, from Wendy, he was a dad who did what he thought was right. In hindsight, you might not think so,' he said.

She nodded but he couldn't see it.

'It's always tempting to judge the previous generation of old fogies by the standards we live by today.'

'Jesus, Johnny, when did you get your philosophy degree?' she asked. But it wasn't harsh mockery, she was smiling.

'I just know that parenthood is humbling. Split decisions come back to bite you on the arse when you never expect it. You know, I read something on Instagram recently that said the best maths you can teach a child is the future cost of their current decisions. If only we'd had lessons like that, eh?'

'You're on Instagram?' She stopped walking and took a breather. She turned to take in the view. It never ceased to overwhelm her with gratitude. London seemed a million miles away, and with it, policing that changed at a hundred miles per hour. In shaky moments she questioned her decision to return to the county she knew like the back of her hand. At times it could be like walking back into a cauldron that one had only just escaped. But she knew that this wasn't the case. This is where she belonged, and he'd brought her up here to show her just that.

She'd walked up here with John Porter hundreds of times. They'd leave Wendy at home, cooking Sunday lunch, and Nikki, her sister, would be out with mates, never one for hiking. They talked about everything: school, friendships, the future, right and wrong. But this is exactly what hurt. She questioned the very foundation upon which he'd built her childhood.

'I joined a few weeks ago and I think my profile picture is outstanding. Sporty but not arrogant. Intriguing without being showy.' He laughed.

'How the hell do you know how to upload a profile picture?' Johnny knew her aversion to social media.

'It's not like Facebook. Posts are all photographic and I only follow accounts that I'm interested in, so my feed is upbeat. I like it, actually. Josie follows me, so it must be pretty cool.'

They carried on after Kelly had given Lizzie an oat bar to suck on and they'd drunk some water. Kelly inhaled deeply and felt her nerves dissipate. She smiled broadly. Johnny caught her arm and kissed her on the lips. The smell of his body, and his touch, had the effect of stilling everything around them.

'Stay there,' he said. He took out his phone and took a selfie of them. Their family.

'I always thought he was a good man,' she said.

'Opinions are all just a collection of feelings towards something,' he said. 'Just because you doubt the past, it doesn't mean your assessment of him was wrong.'

'I'm worried that he didn't protect the people he was supposed to.'

'Because there were kids who got hurt?'

She nodded.

'What if it was out of his hands? The Eighties and Nineties were a different world than the one we live in now.'

'But that's no excuse.'

'I know, but it also hamstrung those who lived through it. They didn't know better,' he said.

'He should have.'

'You can't look back at his mistakes and expect him to have acted against the set of rules he was given then. He was part of a very different system.'

'So that's okay?'

'No, it's not okay, but it's an explanation, and there's nothing you can do about somebody's choices twenty or thirty years ago. He was a product of his time. You were raised differently and you take better care of the people you've sworn to protect. There's something in that.'

'What if I give him the benefit of the doubt and I'm wrong? I'm launching an independent investigation into what went wrong with Brian's case. What if it comes out that he actually screwed up because he was bent. Pure and simple?'

'Then the same applies. You can't change anything. All you can do is make sure those mistakes don't happen again, which is what you're doing. This wasn't your case. It's a historic case, so treat it like that. Okay, so your old fella messed it up. That's not your burden.'

'But if he'd done his job properly, then Brian might still be alive.'

'God, Kelly, you could say that about every single victim on this planet. What if… What if… What is it you always say to me? People do bad shit to good people, and there isn't a cure for that last time I looked. You can do your job, but don't start taking responsibility for those who hurt others, because then you'll go down a hole you can't get out of. If you want to go down that road then why not blame yourself, and your father, for all the ills in society? Where will that get you? And none of it will bring Brian back or change the fact that as a society in general we didn't protect the vulnerable like we do now. You're giving yourself an impossible task.'

They reached the last bit of marshy bog before the summit and had another break. They faced the lake and now, up here, from a couple of thousand feet, it looked quiet and serene, as if the wilderness itself had been watching over the land for millennia, which it had. She told him about what she'd learnt about Steve Thompson's career and the fact that John Porter had endorsed his application for his teaching job at Westmorland.

'It was an old lads club,' Johnny said. 'The army used to be the same, still is in many regiments. The important thing is to weed out the crap when you see it and call people out. I'm presuming that none of the current personnel at Eden House were in position when your dad was? They've long gone?'

She nodded.

'I thought so. Give yourself a break. Everybody makes mistakes, but these aren't your mistakes. And if you're thinking that you should have somehow known all of this when you were seventeen so you could have swung in and saved Brian, that's a fantasy.'

'It's his funeral tomorrow.'

'What time?'

'Midday.'

'So, go and do what you do best. Go and watch everybody who turns up. If your nose is correct on this one, which it always seems to be, then whoever knows the truth will turn up. Somebody who knew him knew the truth. You're not telling me that he got himself into all of this shit without pissing a few people off. Strip it back. Have a good night's sleep, I'll get up with Lizzie, and go with a clear head.'

At the mention of Lizzie, they noticed that she hadn't made a noise for a few minutes, and they checked on her. She was fast asleep. Children were attuned to their surroundings, like most baby animals. Safety and comfort had allowed her to drift off in a haze of security. Others had to fight for their lives.

It made her realise that nothing is ever random. Brian knew he was taking a risk, wanting to expose something that was wrong. There was a part of her that worried that she was wrong about Steve Thompson, even now. After all, he'd kept his job, and his position in the local community. No one had called him out. But her gut told her that she wasn't mistaken. Up here, in the pure air, she knew that, like a cornered animal, Steve Thompson had shrunk into the closest hole he could find, to avoid the detection he'd dreaded his whole career. Now, she just had to find him.

Chapter 49

Kelly was used to funerals. For most people they were the culmination of somebody's life: a celebration of an existence that they were part of. It brought a swell of emotion to the surface that was expelled during the service, and that was the point. It was cathartic. Ancient humans were clever like that. However, for Kelly, since her early career, funerals, except those of her mother and John Porter, were exercises in fact finding.

It was amazing who turned up to pay their respects. The fear of missing out often leads people to make terrible mistakes and they get carried along with the moment, and the pressure of their peers.

The area surrounding the church in the centre of Penrith was rammed and Kelly realised that despite the attempted privacy, word had travelled fast. At least the members of the press were being respectful. She didn't hear shouts and requests for interviews, or calls for soundbites from grieving family members just to sell online later.

They stood in small groups chatting and watching. She instantly knew who they were. They had a way of scouring the surrounding area, like predators, that Kelly had got used to over the years. Anybody who cared about Brian looked grief-stricken. Most people wore black, despite the heat of the midday sun.

Kelly had chosen a black skirt suit herself, aware that her photo might end up in the paper, and wanting to be as anonymous as she could. Brian's peers may look like middle-aged adults now, but when he'd gone missing, he'd been a teenager. A child. A minor, who'd yet to learn who he could fully trust. Like all teenagers, he'd thought he'd known better than everyone else. But like many young people who underestimate danger, he'd paid with his life.

The hearse carrying the coffin arrived and a hush fell along the street. Kelly reckoned there must be 300 people there. One single black limousine followed the main car and it contained Donald and Maureen Miller. A few people made signs of the cross when the cars stopped and the funeral director got out to open the back, making ready for the pallbearers to manoeuvre the coffin onto their shoulders. Kelly looked at the flower tribute on top of Brian's coffin. It was an arrangement of white roses, woven and displayed in three letters, simply spelling one word: son. She swallowed and caught herself. She wasn't here to mourn. Brian had been her peer, not her close friend, even if she did fantasise about being one from afar.

Unexpectedly, as Maureen and Donald were helped out of their car, they spotted Kelly, and Donald walked towards her. He stood before her and thanked her for coming, and Maureen smiled over to her. Kelly composed herself as they went back to the coffin and watched as their son was taken on the shoulders of four strong volunteers. Kelly tried to get an image to leave her head; it was Brian, crumpled up on the mortuary slab, reduced to ginger-coloured bones that looked as though they could have been pulled from a 3,000-year-old bog. His skeleton had told them so much already, but they still hadn't told them who killed him.

Funerals were highly personal affairs and at this one, the congregation was asked to wait outside and follow Brian in to his final resting place. Kelly spotted Tracey and Carol, who avoided her eyes. Then she saw Paul, who waved at her warmly. Then she saw Michelle, who cast her gaze down and wore a net over her face. Kelly thought it melodramatic. She watched Michelle keenly. Brian's parents walked behind him and the rest of them followed in no particular order, but Kelly tried to keep as close as she could to those who might be privy to Brian's last secrets, without causing too much of a disturbance.

There was no sign of Steve Thompson. But Mrs Gooch was there, and Kelly nodded to her.

Several times, she'd woken in the middle of the night and called Eden House to check progress. There'd been no news all night. Now she yawned, conscious that her lack of sleep could come across as police nonchalance.

She followed the pack and got into the church, making her way to the nearest available pew, well aware that the volume of people outside wouldn't fit in. Elton John's 'Candle in the Wind' drifted out of a speaker and Kelly closed her eyes. To her, this was the worst bit of saying goodbye. The severance of the connection to the person through a series of blows to the memory, which stir the vat of feelings exposed by funerals: music, faces and flowers. It was an odd choice of music, but then Kelly realised that when kids die, they haven't left an extensive library of favourite tunes, just the ones they knew in their lives cut short. Donald and Maureen had chosen the most popular song of 1997, thanks to the death of Princess Diana, and Kelly wondered if Brian might have been chuffed or embarrassed.

She heard sniffs and coughs. The church was cool, and a welcome relief to the air outside. Those old medieval engineers knew what they were doing, she thought.

The music came to a stop, and Brian was deposited onto the catafalque before the altar. This was the beginning of the end for Kelly. There was no turning back, or pretending that it wasn't happening. Maureen Miller had to be supported by Donald, even when they were told to sit.

The service melted into a mild hum of music and readings, with Kelly concentrating on those around her.

Her eyes settled on Michelle, who was seated along the same row as she was. The pews were at an angle and so she could see the people to her left and right. It was an arrangement of comfort and very modern, like the church's interior, which was bland, bare and devoid of the pomp of the early church. Kelly always thought it a shame, though it didn't much matter to the dead. Michelle was following her order of service intently, and Kelly saw her hand move to her throat area, pick something off her bare neck, look down, as if checking herself, when she realised whatever it was she was looking for wasn't there, and then touch her ear nervously.

Their eyes met as Michelle looked up. Kelly looked away first.

Chapter 50

Kelly was walking back to her car when she got the news from Eden House that a warrant had been approved and signed by a judge to search the property left to Steve Thompson by his parents. He'd ignored appeals from police, and dozens of attempts to contact him for a good twenty-four hours now, and the judge had agreed with Kelly that it was imperative they find him. It might be that he'd had an accident in his house, and was lying at the bottom of the stairs.

Kelly doubted it.

As she rushed back to Eden House to change into casual clothes she'd packed into a bag, she couldn't help but believe that Steve Thompson had gone AWOL on purpose. Should he not be at the address, and she was dearly hoping he would be, then her next step would be a national warrant for his arrest and a notification procedure to all British ports. She hoped it wouldn't come to that because she desperately wanted to hear his side of the whole story. A tiny bit of her wanted him to tell her it had all been a huge mistake and the gossip surrounding his career was just that: idle chit-chat designed to bring down those who make tough decisions. Equally, she wanted Mrs Gooch to be right too. Mr Thompson could have just been one of those persons in authority who use unpopular methods, and Johnny was right, the Nineties was an alien era compared to now. Corporal punishment had only been outlawed in schools in the Eighties, and teachers were still learning new techniques that didn't involve violence and intimidation, which is what their profession had often relied upon, before they were told, overnight, they couldn't behave like that anymore.

But for the small percentage of her brain wringing these thoughts through her mind, there was a larger, more powerful force telling her to stick to the facts, and dismiss any compassion that she might be feeling due to her rose-tinted vision of the past.

She changed quickly in the ladies' toilets, and went to collect the signed warrant from her printer. Rob was at his desk and he looked as though he hadn't changed clothes since she saw him yesterday.

'Hey, have you slept?' she asked him.

'Erm, I did, gov. I thought it was a bit late to disturb Dan, and I couldn't impose on Kate, so I crashed here. Do you mind?' he asked.

'No, not at all, that's not what I mean. I'd prefer you to sleep in a bed, obviously, but for your comfort, not mine. I'm just concerned you're not looking after yourself, Rob. You're working crazy hours. I'm worried about you,' she said, perching on the side of his desk.

He had red rings around his eyes and they were bloodshot.

'Do you need a break?' she asked him.

'I'm happier being busy.'

Kelly understood how throwing oneself into work might provide a band aid for a period, but it would fall off eventually, and if Rob pushed his health to the limit then it might be disastrous in the longer term. She had a decision to make, and she got the impression she'd have to do it sooner rather than later.

'You look as though you're going somewhere,' he said. 'Can I come? I'd like to get out, I think it'd do me good,' he added.

'Perfect. Let's go. I've got the warrant for Steve Thompson's place. Three squad cars are on their way and we're meeting them there,' she said.

'Let's go,' he said, jumping up and grabbing a deodorant from his desk drawer. She watched him spray his armpits and pop a piece of chewing gum into his mouth. They walked out of the office and Kelly checked her Toughpad as they climbed into the back of a squad car that would take them to the house left to Steve Thompson by his parents.

In the event of the house being vacant, the warrant allowed the police to use forced entry under Section 17 of PACE. The vehicles heading to the property, close to the River Cocker near Brackenthwaite, contained officers wearing public order kit, including helmets and full body protection, including stab vests. They were armed with forced entry kits comprising ram bars, chisels, lock breakers and cutting claws. If they had to, and Kelly gave the order, they'd also use the big red key: a large red battering ram, otherwise known as the enforcer. It was big, it was red, and it opened doors. Kelly had no reason to expect resistance, and even less that they would encounter armed criminal activity. That required another level of entry altogether, and Kelly didn't deem it necessary at this juncture. Part of her secretly still hoped that Steve Thompson was simply inattentive to the local news, and had no idea that he was wanted for questioning in connection with an ongoing inquiry, as well as accusations of historical misconduct. He'd said to Kelly himself that he'd be available should she need him. But she also couldn't escape the creeping reality that no one could find him.

Nerves were strained in situations like this and Kelly was both pensive and dismayed that this was what she was doing with her weekend. She hoped they'd find Mr Thompson plugged into an audiobook, oblivious to the crisis he was contributing to, and this would all be over in a matter of hours, after he explained himself and answered some pretty awkward questions.

They took the A66 out of Penrith and were soon passing Keswick, taking the ring road north of the town to the B-road that linked Braithwaite with Cockermouth. As they sped past the turning for Portinscale, Kelly saw tourists stop and stare as four police cars in formation drove by at speed. They'd have enjoyed a day on Derwent Water, no doubt, sailing, kayaking or paddle boarding. They stared and pointed and children smiled at them, excited at the spectacle. Brackenthwaite was reached from the north, via the Whinlatter Pass, through Lorton, and the road narrowed as the peaks of Grisedale Pike, White Side and Grasmoor shone in the afternoon sun to their left. They were views that were imprinted on Kelly's memory from her childhood. It was spectacular but quiet at the fringes of the national park, and the peaks seemed to nod their indifference to the circus of tiny cars speeding through the valley.

The lead car signalled to turn right, and headed off onto an untarmacked road, which Kelly knew led to the address. It was simply stunning out this way, but Kelly knew that given what Dave Crawley had insinuated, it would also be the perfect location to grease the wheels of a human trafficking operation. Most of the time, you couldn't trust a con, but occasionally they told the truth, just because. They might be bored, they might feel lucky and believe themselves likely to be rewarded with time off their sentence, or they could simply have an axe to grind. The thing about Kelly's job was that every lead had to be followed, whether it turned out to be a pile of sheep shit or not.

The lead car slowed and they followed it to an opening which revealed the house itself. It was bigger than Kelly imagined. It was a grand farmhouse-looking structure, symmetrical and built out of local stone, and she wondered how on earth Bobbie Thompson had been able to afford it. That particular paper trail would have to wait. She told herself that he might have got lucky, or saved his pension, but it didn't fit the profile of somebody who'd been kicked off the force. Honourable or not, alcoholics rarely saved money.

They stopped the cars and got out, checking their equipment. A car matching the description of the one driven by Steve Thompson sat parked to the side of the residence. Kelly and Rob walked to the front door and knocked. The uniform officers awaited instructions.

There was no answer.

Kelly walked around the back of the premises and Rob followed her. From first impressions, they could see that the property was lived in. The log store was full, and a line of dry washing was flapping in the gentle breeze. To the rear, a well-tended and pretty garden was decked out with chairs and a working water feature which consisted of a pool and a couple of statue fountains. Rob knocked on the back door and peered through the windows.

Still no answer.

Kelly didn't enjoy giving the order to force entry in situations like this one. When it was suspected that harm was being used on a vulnerable person, or a crime was being committed on site, or that illegal items could potentially be destroyed as evidence if they didn't enter, then that was different; she had no hesitation then. But this was a quiet residence, and no one was home. Equally, she knew that a broken door frame was nothing compared to what they might potentially find inside, and the odds were stacked in the favour of forced entry. She went back to the front and gave the order.

The uniformed officers filed in order depending on their job, and equipment was passed to the officer in charge of the crowbar, which they'd try first. Contrary to public belief, they did try to contain the damage they caused, if they could. It was only in emergency situations that they literally barged their way in. The crowbar didn't work. The door was modern and made of tough plastic. The big red key was next. They all stood back and allowed the officer to use the huge tool. He swung it back and slammed it into the door, which opened on the third crash. The uniformed officers went in first and shouted for the owner.

Kelly and Rob followed.

The first thing that Kelly noticed was the stifled air. Whoever had left last must have closed all the windows, and in weather like they were experiencing, it wouldn't take long for a home to turn quickly to a furnace. Then she caught a whiff of the stench that every police officer comes across at least once in their career. Kelly had smelled it more times than she cared to recall.

She looked at Rob, who covered his mouth.

Then one of the uniformed officers came down the stairs, coughing and looking terribly pale. He went directly outside.

'Ma'am!' an officer shouted from upstairs.

Rob and Kelly ran up there to where the sound came from.

Chapter 51

Steve Thompson lay face down on his bed. He was naked.

Kelly and Rob stopped at the door.

'Bang goes my witness,' she whispered. 'Fuck.'

Kelly knew that she shouldn't open the windows, which could compromise the scene, but the air was stifling. The temperature in there must have been thirty degrees. She heard the familiar buzz of flies but she didn't see any, then she spotted one landing on the corpse. She took a pair of gloves out of her bag and passed some to Rob, then went to Steve and placed her fingers over his right wrist and felt for a pulse. As expected, there was none. His body was pale and warm, and Kelly knew that it would have settled at room temperature, having dropped slightly from the thirty-six degrees inside the cadaver, as soon as he stopped breathing. It would make establishing time of death complex. The last time he'd been seen was by Mrs Gooch on Thursday as he left work. He'd called in sick yesterday, so he could have been like this for forty-eight hours. She knew enough about corpses to know that in that time the bacteria inside his body would have begun feeding very quickly, especially with it being so warm, and she looked closely at the sides of his torso. Sure enough, there was a green shadow forming across his stomach and out past his ribs, meaning that inside was like a slow cooker. She touched his fingers and they were malleable, meaning that rigor had come and gone, again matching with death occurring sometime yesterday, but she'd leave it to the experts.

They walked around the corpse and Kelly noticed what Rob had seconds before. A plastic syringe lay on the bed next to him, along with a strap and various paraphernalia indicative of serious drug taking. Liquid residue remained in the bottom of the receptacle discarded beside him, and it was the colour of heroin when it melts. It was like somebody had poured a little water in a dirty ashtray.

Kelly had seen plenty of drug-related deaths and had witnessed the sorry sight of the victim in their final resting place. It was always a pathetic scene. For a human being to descend to such a chaotic place, where their only way out is through stupor, was a tragedy that seemed such a waste of a human life. The lack of dignity displayed by the body after it fell into despair was a reminder that this person had lost all sense of reason.

She sighed.

'I'm going to get some air and call forensics,' she said.

When she got outside, and gulped mouthfuls of fresh oxygen, the coppers who'd accompanied her were standing by their cars silently.

'All right, lads?' she asked.

They nodded, apart from the one who'd been sick, and was now sitting inside a squad car with his head in his hands; chances were this was his first dead body. She used one of their radios to report the incident and call for a crime scene investigator and forensic specialists to attend as soon as they could. The dead man was a prime suspect and his death unexplained, and she was taking no chances.

Meanwhile, she and Rob would search the house.

'I'll need one pair to secure the entrance and keep any visitors away,' she said. Two coppers nodded and volunteered for the job; the others, including the young man who was still recovering and sipping water, prepared to leave. Before they did, she retrieved a bag of kit from the boot of the car she'd travelled in. It contained spare gloves, shoe coverings, overalls, masks and evidence bags.

Rob came outside as the other cars left, slowly now, taking care to navigate the dry dusty road. They disappeared around the bushes and she turned to Rob. 'Ready?' she asked. He nodded. They put overalls over their clothes and slipped on masks and hairnets, and went inside.

The place was a tip. The first thing that struck Kelly was that it reminded her of an all-male shared house during her university years. The carpet was stained and unclean, the windows were grey and dusty, the furnishings were dated and old, and the kitchen was a mess.

But they weren't looking to tidy up for the unfortunate victim, they were searching for anything that might explain why Steve Thompson had – it would appear – taken his own life. Of course, it could be an overdose. Even the most seasoned drug users scored bad shit, or injected a little too much, and then there was the possibility that he'd been helped on his way. It was fairly easy to fake a drug overdose, if you knew what you were doing. For the time being, Kelly was treating it as a suspicious death because he was a suspect in a homicide investigation. Until she had confirmation as to how he died, he'd remain so.

They ignored the cosmetic disorder and opened drawers, examined bills and photos and searched through what was left behind of Steve Thompson's life. Kelly had learnt from Mrs Gooch that Steve had trained as a teacher back in 1987. Since then he'd worked in one school, the Westmorland Comprehensive. There was no doubt that he was institutionalised. He also had a questionable professional record. But was this enough to drive him to suicide? The loss of both parents might perhaps have pushed him over the edge, but that was over twenty years ago. The shadow of what this place had been used for since then put Kelly on edge. It had never appeared in the original investigation into the Crawley family trafficking, and so what happened to it? Were there girls and women held here then? Where did they go?

She felt as though she were sifting through thirty-odd years of shadows.

She opened a desk bureau and a fly landed on her hand. She shook it off irritably and imagined the millions of microscopic bacteria it had on its feet after landing on Steve's body, looking for somewhere to lay its eggs.

It didn't take long to establish that Steve had created a mausoleum. There were papers, letters and photos dating back decades. She examined one and turned it over; it said 'Bobbie and Marjorie, 1967'. It wasn't black and white, but sepia mixed with a splash of fading colour, like she'd seen in her own parents' old albums. John Porter had trained with this man in 1969, two years after the photo had been taken. Part of her felt as though she were treading on the history of people who were just ashes and it felt disrespectful. Steve had clearly never let go of the past. She knew from studying profiling that people can get stuck inside a certain period of their lives, if an event or some kind of trauma keeps them there. They never break free, as if time was paused in that moment. Looking around the living room, it appeared that Steve Thompson's life had been paused many years ago.

In another room, Steve's teaching aids were stacked up along a wall. This had been his study. His computer was switched off and silent, and he'd been working on a PE lesson, she saw by the notes left in front of it. She peered closer, because the notes were scribbled so they were almost unrecognisable. The handwriting of teachers, like doctors, was notoriously terrible. There were diagrams too. She stayed there, reading the notes, and found herself sitting down on the swivel chair. When Rob came in and disturbed her, she jumped.

'Jesus,' she said.

'Sorry. There's something you need to see,' he said. 'The forensic van just arrived.'

'Is that what I need to see?' she asked him.

He smiled. 'Very funny. No, it's a bedroom wardrobe.'

'These notes are weird,' she said, getting up.

'What do you mean?' he asked.

'I know he taught PE, but this is more like anatomy. And there's a whole pad of notes on decomposition.'

'Come and look what we found,' he said.

She followed him upstairs and into a room that looked as though it was used as a dumping ground for washing, papers, books, and other last-minute considerations. Kelly knew it could take the forensic team a week to go through it all.

'Here,' Rob said. He looked stern.

Kelly peered into the wardrobe as Rob held the door open.

Hung inside, covered in dust, alongside what appeared to be a woman's dresses – possibly Steve's mother's – was a brown leather jacket. Kelly felt her ears throb and blood drain from her face.

'You are fucking kidding me,' she said to Steve.

'Look at the sleeves,' he told her.

She touched the jacket with her gloved fingers and picked up the end of a sleeve. The leather was the right colour. The design was the right era. The vision of Brian's jacket was seared onto her brain and she could never forget what it looked like. Or Donald's face when he showed her Brian's photo, with him wearing it. It was stiff, like a corpse, and it creaked when she moved it. She pulled out the sleeve and peered at the end of it, where the cuff should have been.

It had been cut off and ribbons of material hung jaggedly.

She pulled out the other sleeve, and found the same damage on the other side.

Chapter 52

Neil Graham packed up the car and forced in the last few items. Why was it that when the kids had been through it, looking for footballs and dolls, the boot seemed smaller? He was angry.

Livid because the holiday was screwed, and irate because they were so far away from home. Of course he did all the driving, his wife got headaches if she went more than half an hour behind the wheel, just enough time to get to the gym, or have coffee with a friend. He was tired and stressed. He'd taken the time off because his wife had nagged him to within an inch of his life. The lure of fresh air and the photographs in the books of the beautiful serene Lake District had tempted him and he'd caved in. Never again.

And that didn't even get him started on the woman who owned the place. Michelle Parkinson played a good game, and talked the talk, but Neil had come across her sort before.

I mean, what sort of a woman goes for a leisurely swim in the lake where police have been searching for dead bodies? His wife told him they were actually searching for a murder weapon, but that wasn't the point. The woman was weird and arrogant. She'd shown no emotion whatsoever since a dead man had been found on her property and he'd read in the paper that she'd actually gone to school with him, as had the copper investigating the case. That had been the last straw. This place was full of incestuous weirdos, he'd decided.

The kids moaned about leaving, but then they complained whatever they did. He couldn't do anything right. He'd lasted four nights and it had been an unmitigated disaster. Ben had been morose and withdrawn, and his nightmares woke them all up. He understood that the boy had seen a dead body, but for heaven's sake, it had been fully clothed and he hadn't gone anywhere near it.

He slammed the boot closed and looked over to the lake.

The owner was at it again, and he could hear her singing. Singing!

He walked to the lake and peered across it. The police hadn't been back since they sent divers in there yesterday. He'd spoken to one of them who'd made polite conversation, confirming it was cold and dark in there. Why anyone would want to swim there was a puzzle to him, but he'd seen it in Derwent Water too. Kayakers, paddle boarders and wild swimmers. They were a hardy lot up here; that was for sure.

He saw that Michelle Parkinson was in a kayak; she was paddling towards the other side of the small stretch of water. He supposed that's how she kept fit, and part of him softened and admired her hard work. He watched as she went close to the bank and stopped the boat and hung her hands over the side. Maybe she was some champion kayaker and this is where she trained? He'd seen her perform this manoeuvre before. She felt the water, as if testing it, then flipped over her boat and paddled out, bobbing up occasionally to try and flip the boat the right way up. But this time she looked as though she was struggling. Neil walked closer to the water's edge and felt as though he had a split-second decision to make. He stripped off his shorts and T-shirt, kicked off his flip-flops, and ran to the water, wading in. Jesus, it was cold. He thought no more of it and dived under, swimming out confidently to where Michelle was. As he got closer, he looked up to check that she was still above water, but she wasn't, and he swam harder and faster. He reached the kayak and looked around it, lifting it up and flipping it over easily.

Suddenly the water broke behind him and he spun around, letting go of the boat.

'Are you all right?' he panted. 'My God, you went under, I thought you were—'

277

'Get out!' she screamed at him.

Neil trod water and tried to figure out what he'd done wrong.

'The sand martins! You'll scare them to death! Get the fuck out of the water, why the hell are you in here?' she shouted.

Her eyes were murderous and her face crumpled in anger.

He kicked the water underneath him and looked around. He guessed that sand martins were birds but he had no idea what they looked like. 'There's no birds here,' he said, staying away from her. She looked as though she wanted to throttle him.

'Neil!' he heard a voice shout behind them. They both turned and saw his wife on the bank, shortly followed by his children. He waved and shouted. 'I'm okay! I'm coming.' He swam away from Michelle, leaving her to whatever she was up to, and made it back to the shore in a few minutes. He reached shallow water and put his feet down, and slipped a little in the mud, which he hadn't noticed when he'd dived in. He managed to walk up the bank and was surrounded by his family, all asking questions at the same time. For the first time in weeks, he enjoyed clarity and the tension in his body had slipped away. He'd found the swim rather invigorating. He smiled broadly at them and his wife looked nervous.

He breathed heavily from his exertion and realised that the physical activity had drained his anxiety level to almost nothing. He'd never felt so alive.

'I dived in because I thought Michelle was in trouble,' he said, and laughed at himself. 'Stupid old fool. Of course she's okay, and she's mad as hell at me because I disturbed her birds.'

'Birds?'

'Nesting sand martins apparently, they live over there by the reeds, on the water,' he said.

'Sand martins don't live on water, though they do like to be near it; they nest in tunnels, that's why they're called sand martins.'

Neil and his wife turned to their fifteen-year-old daughter and stared at her.

'Did you see any, Dad?' she asked.

'No,' he replied.

'Because she's lying.'

'Don't say that,' Mrs Graham admonished her.

'But she is. Lying is when you cover up something with a fabrication.'

Ellie was autistic and struggled particularly with people – usually adults – who ignored factual thinking and tried to see the metaphoric meaning behind everything, which teachers tended to do on a day-to-day basis. It drove her nervous system crazy but also got her into a lot of trouble. She focused on the exact meaning of words, and became incredibly frustrated when adults didn't listen to her.

'You are right, Ellie,' Mrs Graham soothed. 'But a liar is somebody who doesn't tell the truth on purpose,' she added.

'She *has* done it on purpose, to keep Dad away from the reeds.'

Neil looked at his wife and back to the lake. Michelle had retrieved the kayak and was paddling back.

'Come on, everybody, drama's over, who wants to go swimming at Derwent? I've changed my mind, let's stay for another weekend,' Neil said.

His wife stared at his back as he strode away, picking up his clothes, followed by the boys. Ellie hung back and took her hand.

'She lied about the dogs following Ben too,' Ellie told her.

'What, darling?'

'When Ben found the man under the cabin. The dead man with his head smashed in.'

To anyone listening to mother and daughter, it might easily come across as an odd, slightly disturbing conversation, but Ellie was simply stating facts.

'What about the man?'

'It's not about the man,' Ellie said, becoming frustrated.

'Sorry.'

'It's about the dogs, Thelma and Louise. On Wednesday morning, when you and Dad were talking to the lady, and Ben wandered off. I followed him. She came out of the office when you were signing papers and when she saw where Ben was going, she let Thelma and Louise off their leads, from the office.' Ellie pointed. 'She's a liar.'

'Ellie, I've had enough of this, let's go and get our swimming costumes.'

'Mummy! When somebody makes untrue statements, or they withhold facts, then they are lying. Lying is a verb, you have to do it. It's a choice, not an accident. An action, not a reaction.'

Ellie's grip tightened.

'You're absolutely right, Ellie, and I agree with you. She has chosen to tell lies.'

They walked back to their cabin to pack for a day on the water, because Dad had suddenly decided to stay.

Chapter 53

Kelly placed the plastic bag containing the leather jacket on the counter, and Rob made them coffee. She dialled the number for the Cambridge University lab that was currently examining the slivers of leather they'd retrieved off Brian's arms. They were looking for DNA – a long shot after all this time underwater – but now she wanted to ask them if they could look at the jacket itself, something which she never thought she'd be in the position to offer.

She held the photo of Brian in her left hand as she fiddled with the pad and paper in front of her with her right. She'd been unsure as to whether the jacket might be his, but when they'd taken it out of the wardrobe to bag, she'd seen the emblem of Liverpool FC that Brian had sewn on himself. The crest of the Kopites was faded, like the rest of the jacket, but to her, it was like a sparkling jewel in a crown. It jumped out at her. The jacket that Jason had so wanted to wear, but had been ruined by cutting off the sleeves so that Brian could be tied up. It was a depressing vision. Brian must have been terrified.

The lab technician on the end of the phone was positive they could give her a definitive answer to her question should they receive the jacket by courier tomorrow.

'And the fibres retrieved from the car? I believe they were sent to you from Aberdeen?' she asked. The team that had examined Jason Cooper's old Ford had sent their findings directly to her and she'd suggested sending the fibres found snagged under the rear footwell to the same lab that was working on the restraints around Brian's wrists.

'Yes, received. We've been working on those too. In fact, I've got some preliminary results on that for you.'

She waited as the technician moved paper around and she was glad she wasn't the only one who still relied on writing things down, and not just inputting data into a computer.

'It's a mix of organic and synthetic material consistent with the fibres from the leather samples you sent us,' she said.

'The restraints?'

'Yes. Also, the fibres were cut by an object like scissors, rather than a knife. With a knife there would be a sawing action, but under the microscope, it's clear the edges of the fibres were sliced, likely in one motion, consistent with the use of something like scissors. The same is true of the edges of the restraints. As far as extracting DNA goes, we've found organic matter, but now it has to go through a lengthy process of chemical reactions to see if we can gather enough material to work out if it's human, and then see if we can get a profile from it.'

'So, in your opinion, the restraints and the fibres in the front of the Ford come from the same item of clothing?'

'I can't say that, but the same roll of material, for sure. If we find now, after you send us the jacket, that it's a match, then, yes, you have your answer.'

'Thank you. I've never met you, but I could kiss you.'

'That's nice. We don't always get the results we want, but sometimes we do, and it's nice to hear, when you've come into work on a Saturday. I'll be here in the morning if you get the jacket to us.'

The technician told Kelly the name of the forensic courier they preferred to use and Kelly wrote it down.

They hung up. She sat back and closed her eyes.

'Good or bad news, boss?' Rob asked.

'Good, Rob. Very good.'

She told him what she'd learnt and he whistled.

'Why do I feel depressed that all the evidence points to Steve Thompson being our prime suspect?' she asked.

'Probably because it's not supposed to happen, is it? It's like going against the natural order of things. Parents aren't supposed to kill babies, and teachers aren't supposed to kill kids. But there's one thing that's certain now about this case, we can kiss goodbye to an arrest or a trial.'

Kelly nodded and sipped her coffee. They walked next door to the incident room and updated the information that they had on the whiteboard on Brian's half. It was still split in two – Jason one side, and Brian the other. The uniforms who were at work gathered round.

'The forensic results on Steve Thompson's house might take weeks. With the amount of stuff in there which looked decades old, I'm hoping to find some corroborative evidence. But the jacket is what clinches it. I think it's fair to say that we can now piece together a timeline of Brian's last night alive,' she said.

'The next question is if we think Michelle or Jason knew what Thompson was planning. Do we think it was premeditated?' Rob asked.

'I'll pay her another visit on my way home,' Kelly said. 'Without letting her know that we have his jacket.'

Rob bit into a nut bar and carried on. Kelly watched him. When he was stood in front of the board, presenting findings to her team, he was relaxed, focused and alert. She noticed that some of the weight of the world had lifted since he'd told her he was getting a divorce.

Kate came in. 'Afternoon all,' she said.

'Ah, DS Umshaw, what a lovely surprise. We have news,' Kelly said, poking fun at the formality of working on a Saturday when the rest of civilisation was at the pub.

'So I see. My girls went out and changed their minds about spending any time with their mam. I was home alone so I thought I'd come here and join the party.'

'Steve Thompson is dead,' Kelly said.

'Jesus, I should take the morning off more often.'

'We think it's suicide but I'm not holding my breath. We found Brian's leather jacket hanging up in his wardrobe.'

'Bloody hell.'

'Carry on, Rob,' Kelly said.

Kate sat down with the others and Rob brought up a photo of Jason's red Ford Fiesta.

'We think Brian was driven to Thirlmere in this vehicle, and the fibres in the front of the car match the ones tied around Brian's wrists. It's likely they were cut from the jacket and it's just a matter of waiting to get this verified. At some point after eleven p.m. and before midnight, we think Brian was driven to the reservoir, which Steve Thompson knew was drained for maintenance. We think from the lack of further injury to Brian's skeleton that he was incapacitated from the single blow to his head from a spade, either before or after he was taken there. Then he was stripped and buried. For some reason, Steve Thompson took the jacket home.'

'Why do you think they left the socks on?' a junior officer asked.

'Good question,' Kate said.

'I spoke to Henry Dempsy about this. He reckons that they stripped his clothes, partly so he couldn't be identified but also because it would make him less likely to float and be spotted if the grave didn't hold once the water was pumped back in. They didn't think the socks were important, or they were disturbed and ran out of time. I'm curious, though, as to why the jacket was destroyed for this purpose. Joanne Cooper said Jason coveted the jacket and desperately wanted it,' Kelly said.

'Maybe it was done on purpose so that he would never be tempted to wear it,' the officer said.

'Good point. It's extra circumstantial for us that Jason was involved,' Kelly said. 'Now we wait for the forensic results from Thompson's place and the autopsy to prove or disprove that he committed suicide because of the guilt.'

'The guilt of doing it, or the shame of getting caught?' Kate asked.

'I think we know the answer to that.'

'It still doesn't tell us who killed Jason. The lake search found nothing, and they were able to cover a fair amount of area.'

'That reminds me, boss,' Rob said. 'There was a message from a guy called Neil Graham this morning. I've just picked it up. He wants somebody to get back to him about the lake.'

'The lake? Graham, I know that name, it's the father of the boy who found Jason's body. Right. Is he still at Parkie's?' Kelly asked.

'I'm assuming so. His message said they'd be back after visiting Derwent this afternoon.'

'I'll get off then,' Kelly said. 'Wrap up what you're doing. We'll have to wait for lab results until Monday. Rob, can you courier the jacket through this company to the lab in Cambridge for me?'

'Sure,' he said.

The group dissipated. Gone were the days of unlimited overtime; now there were a certain amount of extra shifts you could work and so they all had to be gone by five p.m. or they wouldn't get paid.

'Kate, can you go through the prelim results for Jason's phone?'

'Sure. Rob, I've got a casserole on for us later. You're staying at mine tonight,' Kate said.

'I heard that, no excuses for staying on the office sofa tonight,' Kelly said to him.

She left them working together.

As she exited the rear of the building, a sea of reporters shouted at her from the barrier. They asked about Steve Thompson. He had no next of kin, and the information of his passing had been kept to a small group of people. But they knew about the activity surrounding his address, and the fact that forensic vans had been arriving all afternoon. It also wasn't breaking news that the man hadn't been seen since Thursday. The journos had put two and two together, as they always did. She spotted the female journalist who'd shoved a mic into her face on Tuesday asking about the skeleton in Thirlmere. She'd covered Brian's funeral this morning. It seemed weeks ago now since Brian's body had revealed itself, thanks to freak weather, and a lifetime since Kelly had first gazed upon his bones, not knowing that it was her classmate who'd lay there for almost a quarter of a century.

She silenced the demands from the reporters and got into her car.

Chapter 54

Parkie's was quiet. It wasn't a surprise to Kelly, who knew that on any Saturday during high season, tourists would be out and about, hiking, taking afternoon tea or shopping in Keswick's boutiques. Neil Graham had said he and his family were back from a trip to Derwent. He'd sounded a different man over the phone when she'd returned his call. The abrasion had gone, he'd been open, keen and helpful, and it had come as a shock, especially when he'd said how urgent it was that she meet his daughter.

She drove past the office and followed the wooden arrows to cabin thirty-seven. Sure enough, his large Audi was parked outside and Mrs Graham was heaving suitcases out of the back. Kelly got out and greeted her warmly.

'I didn't think you'd come so quickly,' Mrs Graham said to her. 'You must be busy, I hope we're not wasting your time,' she added.

'Not at all, it sounded important,' Kelly said, removing her sunglasses and peering into the open door of the cabin. 'Is this a good time?' she asked.

Mrs Graham stopped her task and dumped a large holdall at her feet.

'Getting ready to leave?' Kelly asked. She'd informed Mr Graham that their part in her inquiries was done and they were free to move on. She'd guessed that they couldn't wait to put their Lake District holiday behind them.

'Neil changed his mind. We're staying. He went swimming in the lake this morning.'

Kelly looked behind her to where the police tape still flapped around the perimeter of the beautiful body of water and frowned.

'It wasn't intentional,' Mrs Graham spurted. 'He thought the owner was drowning.'

Mr Graham appeared behind his wife and smiled broadly. 'Detective, come in. Thank you so much for coming, can I get you a drink? My daughter is inside.'

Kelly was thoroughly confused. She looked between the two of them. She could do with a cold drink now she was here, she thought. She followed him inside and was aware of Mrs Graham following her.

'You said I should call if I thought of anything at all,' Mr Graham said.

'You listened?' Kelly said.

He turned and smiled at her, putting his hands up in defence.

'I apologise for my mood. It's not every day you go on holiday and your kid finds a body. It's taken me a few days, but these hills have got to me, I admit it.'

Kelly noticed that his shoulders were freer and he wore no footwear. His shirt was open and anyone would think he'd lived here all his life. She was secretly proud that her home county had this effect on people. But she still didn't know why she was here, apart from to talk to his daughter.

'Ellie!' he shouted.

A girl came into the living area and Kelly smiled at her.

'Ellie's autistic,' Mr Graham said.

Kelly winced. It sounded like an apology and an introduction that was unnecessary, especially in front of the child, as if the girl was abnormal in some way. Mr Graham read her disapproval.

'It's important, it's why you're here,' he said.

'Hi Ellie,' Kelly said gently.

'I'll sit here,' the girl said, without responding to Kelly, or looking at her.

'The lady lied,' she added.

Kelly sat close to her, on a wicker chair, and leant over, resting her elbows on her knees.

'Which lady is that?' Kelly asked.

'Ellie is going to tell you in her own words exactly what she told my wife and I this morning, when I went to rescue the park owner from a kayak incident, which turned out to be not an incident at all actually.'

'Dad, you're interrupting. I think I can remember everything on my own, because it's in my head, not yours,' Ellie said.

Kelly looked at Ellie, then to Neil Graham, who shrugged. Then she listened.

Chapter 55

Kelly found Michelle in her office. For once, she wasn't wearing wellington boots, but trainers. She seemed to be finishing an important task when Kelly startled her.

'Kelly, hi, I was just thinking about you. I'll make us a nice coffee.'

Michelle disappeared into the small kitchen and Kelly heard the clink of cups and the pouring of milk into the frother. Kelly glanced around. Before going to find Michelle, she'd sent a message to Kate. Circumstances had changed. And when that happened with a suspect, it usually meant that volatility followed.

'Is this a social visit? You're not working on a Saturday?' Michelle said from the kitchen.

'You didn't see me drive in?' Kelly asked.

'Er… no.' Michelle replied.

Kelly stood in the middle of the office, looking at posters on the walls advertising Keswick and local walks.

Michelle emerged with two mugs of coffee and gave one to Kelly. Michelle's hand was steady, but there was an edge to her, like there'd been when they were seventeen.

'Do you hike?' she asked Kelly.

'Of course, when I get the chance. You?'

'Nah, I don't get the time, I did plenty of that with my old fella,' Michelle said.

'So, it wasn't all bad then.'

'I should have pushed him off a cliff when I had the chance,' Michelle replied.

'Shall we walk? It's a lovely afternoon,' Kelly said.

Michelle shrugged. 'I heard you've only been back a short while. How was London? If I were you, I'd have stayed down there.'

'Well, it was good for a while, you know, different, but then the pull of this place brought me back.'

'You must have been important down there, it can't have been easy making that decision. Some of us were destined to stay, though if things had been different I would have been long gone.'

'What kept you here?'

'All this,' Michelle said.

They went outside with their drinks.

'Do you miss him?' Kelly asked.

'Who, my old fella?'

'No, Jason. You two were pals.'

'Ah, erm, no. He was trouble.'

'That's what his sister said, you kept in touch with Joanne?'

'We picked up the pieces of Jason's life when we needed to. Nice service this morning.'

'Yes, it was. It was good to see so many old faces. Pity Steve Thompson wasn't there, he had great affection for the kids in his care,' Kelly said.

She felt Michelle's stare on the side of her face. It was brutal and raw.

'Have you listened to nothing I've told you?'

The switch in mood was just what Kelly had expected. Buying more time, she steered them towards the lake, past Jason's cabin.

'I was just making the point that I thought he'd be there,' Kelly said.

'Have you found him yet?'

'Yes, we have.'

'And?'

'We're working on the case with him.' It wasn't a lie. Steve Thompson's body, as well as his house, might just help with her inquiries after all.

'Really?' Michelle asked.

They walked towards the lake and Kelly felt the prickles of uncertainty snap at her skin. Walking past Jason's cabin focused her mind and she experienced a moment of clarity. She got her phone out of her pocket. She looked at Michelle and back to the cabin.

'Sorry, it might be important.' She pretended to open a message. Michelle whistled loudly and Kelly almost dropped her phone. Thelma and Louise appeared and sniffed around Kelly's feet. They were inquisitive for sure.

'I thought you kept them tied up,' Kelly said, looking at her phone.

'No way! Not these two. They love wandering about on their own, and they're no trouble.'

Kelly knew from Ellie that the dogs had been tied up last Tuesday morning, and Michelle had let them off when Ben Graham went exploring. She found what she was looking for on her phone and zoomed in. It was a photo of the framed picture that Jason had in his cabin. It had nagged at her for several reasons. Not only because it was taken such a long time ago, and seemed to suggest that Jason's nostalgia for his school days was unusual, but also because the group of friends had since broken up, quite unceremoniously. Why he'd kept it bothered her. But it wasn't that which disturbed her now. She'd studied the photo a few times, examining the faces of the friends, wishing, in a sense, that she was in the photo herself. And that had clouded her judgement of it. She looked at it with fresh eyes and remembered why it had stuck in her head. She zoomed in on Michelle and noted the way that her hand hovered at her throat, just as it had inside the church this morning at Brian's funeral.

Michelle wandered ahead of her and Kelly stopped.

In the photo, Michelle beamed back at her, as did all the faces in the group. Paul, looking young and handsome and full of ambition. Brian, innocent and crazy, as if he was about to surprise them all and goof around. Jason, broad and manly. Carol and Tracey hanging on the periphery, like cheerleaders. And Michelle, holding a silver locket, out of place for a day on the water, which is why Kelly remembered it. The chain caught the light and her hand cupped the piece of jewellery lovingly.

This morning, at the funeral, Michelle had absentmindedly gone to touch it, at her neck, because she wore it always. But she'd lost it, and missed it terribly. The way she looked when she realised that it was no longer there was a mixture of regret and frustration.

Kelly put her phone away.

'I remember your mum,' she said, catching up.

'Really?' Michelle said.

'She was always quiet.'

'Recovering from a beating,' Michelle chipped in.

'You must miss her,' Kelly said. She watched as Michelle peered out to the lake and across to the other side. It was so peaceful out there. Kelly's eyes wandered to the area at the far end that Michelle had told her was protected by the National Trust because there were nesting birds over there. The parameters of the dive search had been based upon this very piece of information, to avoid the birds. They'd worked on the theory that somebody had thrown a heavy yard tool into the water from close to Jason's cabin. The bird colonies were at the furthest point away from the cabin, so it had seemed a waste of time to include that area of the lake.

'Have you got a picture that you keep? I keep one of my mum and dad on my desk at work,' Kelly lied.

'Only my mam. I couldn't look at his face if you paid me. The day we buried him, I went back at night and pissed on his grave,' Michelle said.

'I'd like to see it. Not your dad's grave of course, the picture of your mum,' Kelly said, finishing her coffee.

Michelle smiled and Kelly waited for an age, listening to the breeze and sensing the lake's surface rippling, until Michelle nodded and led the way.

Michelle took her to her cabin.

The interior was pristine and smelled of cleaning products and air freshener. It was nothing like Kelly had expected. The furnishings were modern and bright. The windows were open and a delightful breeze flowed throughout the space. Michelle went to the mantel above the fireplace and took a framed photo off it. She stared lovingly at it and passed it to Kelly.

'Here she is.'

Kelly held it, trying not to let the adrenalin rushing through her blood show in her grip.

'She was beautiful,' Kelly said, staring at the image of Michelle's mother. It was like any other portrait from, she guessed, the late Seventies. Her hair was coiffured and sprayed, her make-up feminine and stern, and she wore a ruffle at her throat. The locket was placed perfectly on top of the netting and Kelly could see its every detail.

'That necklace is the only thing I have left of her,' Michelle said, coming up behind Kelly and taking her by surprise.

'Do you wear it?' Kelly asked, willing the squad car to hurry the hell up.

'All the time.'

Kelly looked at her throat.

'I misplaced it. I'm gutted.'

'I bet you are. It looks very valuable, was it an heirloom? Have you retraced your steps? I mean, did you lose it recently?'

Michelle's face changed and Kelly eyed the door. Michelle stood in the way. The game was up.

'You know how this ends,' Kelly said.

'Fuck you,' Michelle said and took a step closer. Her screwed up face was the same one she wore twenty-four years ago.

'That would be a very stupid thing to do, Michelle.'

They faced each other. Michelle clenched and unclenched her fists. Thelma and Louise began to bark.

'Strong women don't do what you did,' Kelly said. 'Michelle Parkinson, I'm arresting you on suspicion of the murder of Jason Cooper.'

Suddenly Michelle lunged at her and Kelly ducked, squatting to the floor. Michelle darted out of the cabin and Kelly sprang up.

'Shit,' she said. She ran out of the cabin and to her car, grabbing her radio and requesting the attendance of officers to Parkie's cabin park. She spun around and banged straight into Neil Graham. Kelly dropped the radio and sprinted in the direction that Michelle had gone, and thanked herself for wearing trainers. She saw Michelle disappear behind the office and ran around the other side.

Michelle was stocky but fast, and Kelly kept her eyes on her as she negotiated the rough ground behind the park. She saw Michelle climb awkwardly over a fence, and then Neil Graham overtook her on her left side.

'I'll get her,' he said breathlessly.

Kelly sped up. 'Stay away,' she shouted. The last thing she needed was a civilian getting caught up in something they really didn't understand. Michelle could be carrying a weapon.

There was a small hill to negotiate and they ran around opposite sides. Kelly emerged first. They were at the north end of the lake now and the only sound was of their feet pounding the ground.

'I've lost her,' Neil shouted.

Kelly turned back and sprinted up the hill, and stopped at the top to see if she could tell where Michelle had gone.

Kate called her phone and she answered it, breathless.

'Stop! Leave her!' she shouted to Neil, who stopped running and came back to her vantage point. She saw Michelle's figure disappearing into the distance and knew she wouldn't get far.

'Kelly?' Kate asked, panicked.

'Kate, I'm good. Michelle Parkinson has done a runner on me. There's a dirt road to the north of Parkie's, it goes over the river and towards Lonscale Fell, behind the approach to Skiddaw from the east side.'

'I know it.'

'Direct the units there, will you. I'm on foot, she's gone.'

Chapter 56

Within an hour, Parkie's had been turned into an open-air incident room. Neil Graham's daughter had been correct. Sand martins nested in earth, not reeds. Kelly arranged a second search, concentrating on the area at the far end, where Neil Graham had swum to. He had thought that Michelle Parkinson was drowning, when in fact, she was likely checking that her murder weapon was still undetected. Perhaps she planned to move it. Possibly she simply went there to inspect her territory, returning to a place connected to the crime she'd committed.

Whichever scenario was the truth, Kelly was now convinced that Michelle Parkinson had killed Jason Cooper when Brian's body was discovered. Her motive remained unclear but could have two possible explanations: either Michelle had waited all these years for revenge, never quite knowing if Jason had really been involved in Brian's death twenty-four years ago or not; or she'd been involved and didn't trust Jason to keep his mouth shut.

As for Brian's killer, the discovery of the jacket was enough to convict Steve Thompson – however, that would now never happen because his body lay stiff on Ted's mortuary slab.

The dive team, having been told to expect a potential second search, was able to set up quickly, and the sergeant liaised with Kelly. Search and rescue teams had been mobilised to look for Michelle. Kelly had other work to do.

'I'm concerned about the thickness and depth of those reeds,' the dive sergeant said.

'I know,' said Kelly, watching the screen. She stared at the same murky depths as she had only yesterday morning. Virtually the whole area from the park side of the shore out to a radius of fifteen metres had been searched, ignoring the reed beds protected by the National Trust at the other end. But now she'd changed her mind. They watched the live feed from the body cameras placed on the divers' suits. They wore closed-head breathing apparatus which allowed them to talk, and they described the conditions at the bottom.

The reed bed was thick and forest-like, and the diver fed back his trepidation at going in. He wore a lead line and his buddy stayed out of the thick mass, in case he had to perform a speedy recovery in the event of a line getting caught.

The van was parked at the other end of the lake this time, and the view from here back to the park was quite different. Only a criminal with good local knowledge of the area, as well as time and confidence, would walk, or paddle over here to dump a murder weapon. They spoke in whispers because they were close to the nesting sand martins, after gaining advice from the National Trust, who'd tried to block the search going ahead, until it was explained to them that it was part of a criminal investigation, and they had no choice. The tiny sand martins were nonchalant and Kelly was impressed with their hardiness. Of course, Michelle's ruse that they were easily spooked was a myth, and they carried on their daily routine despite the intrusion. They nested in holes in the bank, dry and sandy from the long heatwave, and they flitted in and out above their heads. Kelly had thought them swallows when they swooped and dived with their graceful arrow-like bodies, darting this way and that. Their dark underwing contrasted sharply with their white bodies and their speed was incredible.

'I'm descending to five metres,' the lead diver reported.

She watched as the screen became a dark green and she could just make out the shadows of upward streaks, denoting the direction of the reed growth. It was like an underwater cage.

She watched as bubbles ascended above the divers' heads.

And then something came into view that didn't belong there. It was how she imagined the discovery team that located the *Titanic* in her final repose at the bottom of the Atlantic felt. The diver's camera revealed a sharp edge that shone a tarnished silver in the murk. Kelly's stomach churned with butterflies and she willed the diver on to rush and pull the object out of its tangle. But he took his time. The diver parted the reeds around the shape he'd seen, and pulled ever so gently on the thick tendrils of the roots, to separate the thick green shoots.

A handle.

An edge.

The diver placed his hand on the handle and pulled gently and the thing came away from the reeds easily. It filled the screen in the van and Kelly smiled and slapped the shoulder of the sergeant.

'What you wanted?' he asked.

'Yes, thank you, that'll do nicely. Good job.'

She walked out of the van and went to the water's edge, watching the bubbles from the lead diver break the surface. The body of his buddy bobbed on top and he surfaced with his colleague, who held the spade out of the water. They floated to the shore, having descended well within their limits and stayed in the reed bed until they could be helped ashore. They'd been so graceful underwater and now they needed the aid of their colleagues to unburden them of their heavy kit. Two others waded into the water and took cylinders, weight belts and fins off the pair, another took the spade with gloved hands and placed it into an evidence bag.

Kelly hurried from the van and peered at the weapon underneath the plastic. It looked benign, ideal for garden work and a whole host of chores, not including homicide, but when somebody is intent on causing harm, they'll use anything, and the most innocent every day item can become lethal. She slipped on gloves and took the bag, thanking the team. It was then put into a black outer bag. She made her way back to a waiting squad car, and told them to get it to the Penrith and Lakes mortuary as fast as they could.

Ted Wallis and Dr Dempsy would be expecting them. Then she turned her attention to the search for Michelle Parkinson.

There was a limited number of places she could go, and she'd soon get tired, hungry and disorientated in this heat. The human body could survive without food, and even water, for a time, but it was the demons inside the head that were the biggest drains on resistance. And Kelly wouldn't want to be inside Michelle's head right now.

Chapter 57

The download from Jason's phone was sent to Kelly's inbox as she drove to the Penrith and Lakes, along with Rob's notes on them. She read the notes after she'd parked and saw that the last caller to contact his phone was at five fifteen p.m. on Tuesday afternoon, and it had been Michelle Parkinson's mobile.

That wasn't all. The call before that had been outgoing, and it had been to Steve Thompson's personal mobile. It suggested panic among the threesome.

She walked into Ted's mortuary and greeted Henry, who was sat on a steel chair, chatting to a technician.

'Have you got it?' he asked.

She nodded and proudly presented the spade, found at the bottom of Michelle's lake, to him.

'I'm no expert, and it might be wishful thinking, but the head looks dented to me,' she said, handing the bag to him.

Ted came in from the chill room wearing scrubs and nodded to her.

'I caught you two before the pub beckoned then?' she asked him.

'Only just. You were lucky, I was showing Henry the view from the steamer and we decided to give the bar a miss, opting for afternoon tea at the Glenridding hotel instead.'

She smiled.

'Right, what have we got?' he asked.

Henry unzipped the black bag and revealed the spade inside, beneath the plastic evidence bag. Ted placed it carefully on the gurney beside the bag and Henry set about measuring it and weighing it. He examined the damage, where Kelly had thought she'd seen a dent. Henry peered closer.

'And you say it was a woman?' he asked.

'Yes. A friend of his. Ex-lover actually.'

'Do you think she could wield this thing hard enough to disable a six-foot male?' Henry asked.

'In a rage, yes. I think twenty-four years of hate and the desire for revenge would do it,' she said. 'I don't think Brian's death was real to her until we found his body, then the hate came flooding back.'

'But where does that leave your theory about the teacher being responsible for Brian's death?' Ted asked. 'Why didn't she direct her rage towards him?'

'My guess is that she saw Jason as the one who delivered him and caused the trouble in the first place,' Kelly said.

'That's reasonable doubt, right there,' Ted said.

'But Steve Thompson will never be tried, and besides, you've yet to tell me exactly how he died,' Kelly said.

'Which is why I'm dressed like this,' Ted said, doing a twirl.

The body of Steve Thompson had been delivered to the mortuary after the crime scene investigator had finished up at the house. The forensic team was still there and Kelly reckoned they would be all next week too.

'I'll go and get him,' Ted said. 'Henry, will you assist?'

'Of course,' Henry said, getting off his seat. Henry wasn't a medical doctor, but he understood how tissue adhered to bone, and so Ted valued his opinion when it came to the deeper study of a corpse. To have the specialist sat in his mortuary at his disposal was an opportunity for a supporting opinion, and a second pair of eyes.

Henry went to get overalls on and wash up. Kelly took a seat on a high steel stool overlooking the gurney on which Steve Thompson was to lie, and give his last lesson.

The body was wheeled in and Ted read out the crime scene report.

Ted's job as pathologist was to rule out morphological causes of death, in other words, anything other than a drug overdose, as was written in the report when the body was discovered. He'd try to establish whether suicide, or even homicide, was in play, rather than anything else. Kelly had seen dead bodies after drug overdoses before, but it was fairly easy to construct such a scene, if anybody was that way inclined. The Home Office recommended that all drug-related autopsies were carried out with a healthy dose of suspicion. No pun intended.

'So, no sign of forced entry. Ambient temperature inside the property – with windows and doors closed – thirty degrees, matching temperature of body, taken by paramedic…'

Ted spoke into his mic and Henry came back into the room. 'Body naked, face down…'

'I've brought his medical records,' Kelly said. 'He had high blood pressure, and that was treated with statins. He was diagnosed with anxiety disorder in 2010 and put on citalopram, twenty milligrams daily. He complained of headaches. His alcohol intake according to his GP was high – at forty units a week – so we can double that,' Kelly said, knowing that everybody lies to their GP about routine drinking.

'I'll soon know when I get to his liver,' Ted said.

'Nothing on illicit drug use, but that's normal if he was a recreational user,' Kelly said.

'Sadly, intravenous use is now considered pedestrian,' Ted said.

Kelly had to agree with him. It was possible to keep relative control of your life and still take hard drugs at the weekend, if you were careful.

'ECG in 2019, suspected myocardial infraction but later suggested it was pericarditis,' Kelly read. 'Negative for HIV. He had a test in 2020.'

'He was young to have all that going on,' Ted said. 'But it could all be explained by illicit drug use. Let's have a look.'

Kelly recalled Steve Thompson when he was at his best: teaching. It didn't matter what he was trying to get into his students' heads, he always told a story, and that's one of the reasons why she'd thought he was born for the profession. The things that she'd discovered about him recently had driven a torpedo through all of that, and she watched as his body emerged from the black body bag, the teeth of the zip grating on her nerves.

Each body looked the same in death: empty, vulnerable and still. It seemed trite to describe a corpse in such a way, but animation was one of the things that gave people their personality, and that was one of the things most obvious about cadavers: the evaporation of life, of movement, of spark, of soul.

She had trouble looking at him. She saw, instead, his smile, his eyes when a student answered something correctly, his dive into the lake to rescue Brian, and the sincerity in his face when he'd lied to her last week. As he lay there now, he was none of those things, and his blackened fingers attested to that, where the CSI had taken his fingerprints at her request, so they could compare them to the ones found on Jason's steering wheel. He was a dead criminal.

'Do we know if he was right or left handed?' Ted asked.

'Right,' Kelly said. She surprised herself. She just knew, because she remembered him writing on the board. It was a long hidden memory that her brain surely never thought it would need again, but up it popped.

'Fits with the injection site in his left arm then,' Ted said.

Kelly nodded.

The body was lifted out of the bag and it settled like a great half-tonne marlin being landed on a boat, like on the nature programmes John Porter used to watch.

Ted began his external examination.

He started with the arms and legs, in the lymph regions and glands, looking at the major vein sites. 'I've got several recent and healed needle puncture marks here,' Ted said.

Kelly's heart sank.

Her teacher had been a junkie.

Kelly knew, because she'd found him, that Steve Thompson had also shit himself, another tell-tale sign of overdose. But if he was a chronic user, as his healed puncture holes indicated, then he'd know his limits. But being aware of how much substance one could ingest didn't account for what it was cut with on the street. Suicide in these cases could only really be determined by a series of circumstantial factors, such as the state of the victim's personal life, mental state, finances, the scene of the death itself and the amount by which the individual overdosed, and they wouldn't know that for days until toxicology came back.

When Ted pulled apart Steve Thompson's body to begin his internal exam, as he would be doing soon, then Kelly planned to leave. She'd already established that there was no trauma, because his routine X-ray had been clear, and that he was a regular drug user, with mental health problems. She didn't need to see his liver being cut up. Unless Ted came up with something spectacularly surprising, then this would be ruled as either an accidental overdose or a suicide. The single fact that he was wanted by police could have tipped him over the edge. But Ted's main job was ruling out homicide, because at this stage, knowing what she did about Steve's history, it was just as likely that he had enemies who'd benefit from him never going to trial. Perhaps enemies who were already behind bars. Enemies who behaved like friends. Enemies he sent parcels to.

Kelly needed to find Steve's dealer, and to do that she needed to locate his phone, but it hadn't turned up inside Steve's home. Yet. It was about to become clear why. Ted interrupted her train of thought by asking if the cadaver could be turned. As the technicians did so, a mobile phone was nudged to the side of the body.

'Hello.' Ted said. 'It looks like this was underneath him, can you check the notes?' he asked her.

Kelly scanned the report and found that when the medics had moved the body, he'd been lying on the mobile device and it had been put inside the body bag, as was standard.

'May I?' Kelly asked.

Ted nodded.

She took it with her gloved hand and pressed the home button. It was charged, and the home screen flashed up. 'Excuse me,' Kelly said.

'Be my guest,' Ted told her.

Kelly took Steve Thompson's lifeless thumb and used it to open the device.

It worked.

Now Steve was on his back, and it made Kelly's melancholy worse.

'Bloody foam in nasal cavity,' Ted said. 'As well as powder residue.'

Kelly looked up and peered at Steve Thompson's face. His eyes were open and the skin around them was red raw, as if he'd been crying right up until the moment they found him. His torso was fully green and Kelly got a whiff of putrefaction. Some corpses ended up like this because they were snuffed out at the hands of a madman – or woman – but Steve had done this all himself. He was a shell of a human being, and what was left now was rotten.

She looked away and concentrated on his phone. She searched his call history and noticed Michelle's number appeared several times last night, as incoming and outgoing. She knew that among the last calls he made would be the name of his dealer, because junkies were never discreet. Michelle's number, though, was stuck in her head because not only had Michelle called her phone, with helpful information about the lake – or so Kelly had thought at the time – it had also appeared on Jason's phone report. Then she checked his Google history. It hadn't been cleared and she breathed a sigh of relief, trying not to take in too much of the gasses escaping from what was left of the man who'd made them.

Brian Miller's name was searched fifty-three times.

Parkie's was searched twenty-eight times.

Kelly Porter thirty-nine times.

She clicked on his Facebook page, and his search history. Then his Instagram profile. He'd been looking at posts about Brian's funeral. There were two Facebook pages set up for it, between old school friends, and he'd liked posts from numerous contributors. One was Carol Fisher and another was Tracey Dalton.

She searched through his photo albums as she heard Ted start his electric saw to open up the body. Kelly knew that with a suspected overdose, and a history of alcohol abuse – as verified by Mrs Gooch, for one – Ted would be examining certain organs more closely than others. He'd pay attention to his oesophagus for signs of laceration or retching, his pancreas, which was likely to be enlarged, as well as his liver, for signs of chronic disease. Pretty much every square inch of the body was fucked up by alcohol and hard drugs, and Kelly had seen it with her own eyes too many times. Steve Thompson had looked normal to her on the outside, when she'd seen him earlier in the week, but now she heard Ted talking into his mic and detailing what drugs and alcohol had done to his insides, she reckoned she wouldn't be drinking a glass of red tonight.

She stopped scrolling when she found a screenshot of an original photo of her class on their school trip in 1996. The album contained others like it, which also must have been screenshots of standard photos, given their dates. There were a couple taken from newspapers, and one of Brian in his leather jacket – the one given to the police by Donald Miller – as well as shots of groups of friends. Peers of hers. Groups of kids on the water, from years back. She didn't recognise some of them, but in others she saw her own face beaming back to the camera, off centre, never in the in crowd, but part of the group nonetheless. In every shot of their year group, Jason and Michelle took centre stage.

Kelly almost dropped it when it rang. She looked at the caller's ID, and it was Paul Gordon.

Chapter 58

Kelly answered the phone and remained silent, hoping to encourage the caller into identifying themselves.

'Steve? Steve?'

She said nothing.

'Steve? Are you there? Mate, call me. For God's sake, call me.'

Kelly held her breath, wanting with every fibre of her body to speak. She pressed End Call.

'Everything okay?' Ted asked, bringing her out of the zone of blackness that she'd so easily slipped into. Nothing made sense. Paul had called him 'mate'.

'I have to go, Dad. Call me when you're done.'

'The toxicology will take time,' he said. 'If I find anything interesting in the meantime I'll let you know.'

He stood with his saw in one hand and forceps in the other. The pathologist's toolkit looked like something from a horror movie but she reminded herself that Ted was one of the good guys. Steve's body moved a little from the withdrawal of Ted's instruments and she thought for a second that he might sit up and tell her what the hell was going on.

As she left the mortuary and ripped off her gloves, bagging Steve's phone and placing it in her bag, she called Kate, who was still at Eden House.

'Boss, we've got some interesting phone calls going on between several characters on our list.'

'I can raise you on that one,' Kelly said. 'Steve Thompson's phone was under his body, and it has just received an incoming call from Paul Gordon,' Kelly said. 'Any news on Michelle Parkinson?' she asked.

'We've got a helicopter over the area with heat-seeking equipment on board. Mountain rescue have also sent a team up to Lonscale Fell to search for her.'

'Can you triangulate calls to and from the numbers belonging to Jason Cooper, Paul Gordon, Steve Thompson and Michelle Parkinson?' Kelly asked her.

'I'll get on it now,' Kate said. 'Rob has joined the mountain rescue team, he's on his way to Lonscale now.'

'That's not a good idea, he's exhausted. He shouldn't go.'

'That's exactly what I told him, too. He's gone anyway.'

'Damn. Can you see if you can bring Paul Gordon in for me?'

'Sure thing.'

They hung up.

She texted Johnny. 'Are you on call?' she asked him.

He texted back a thumbs-up emoji and said he'd been called out by the police to Lonscale Fell, and Josie was looking after Lizzie.

'Be careful,' she messaged him. 'Potential murder suspect.'

He took a few minutes to reply with another thumbs-up and scary face. 'Great.'

She left the hospital and walked to the car park, getting into her car and pulling away impatiently. She was dealing with nothing but liars. Paul had lied to her by telling her he'd distanced himself from the group. Steve had lied too, and so had Michelle. She thought about who she could trust to tell the truth, and she was stumped. Who was it who'd said the truth will set you free, but not until it's finished with you? She pulled out into traffic and put her foot down, tempted to use blue lights to get through town.

As she approached Eden House, she nudged her car through the cordon of journalists waiting to hear any news on either murder. Nationally, the interest had died down, as it tended to,

309

as other news took over, but locally, the vultures still circled. She parked and went through the back entrance, taking the stairs. She found Kate in the incident room going over phone numbers and logging a pattern of calls, which had intensified since Tuesday morning.

It was Brian's parting shot, Kelly thought to herself with satisfaction. He was framing these fuckers by emerging from the mud. They'd panicked and begun planning what they were going to do, but none of them stuck to the programme, that was plain enough to see.

'Boss,' Kate said. 'Paul Gordon has been picked up at home. He didn't want to come in but he didn't want to be arrested either.' Kate smiled.

'Good, let's get a few interview suites ready, it's going to be a long night and we need to get our timelines spot on.'

'I've set up a live feed linked to Rob's bodycam,' Kate said.

Kelly watched on a screen to the left of the whiteboard as Rob's camera went live. Her only option now was to put pressure on those who were still alive to tell their story. Groups of people, once so close, often cracked and split on each other very quickly when interviewed simultaneously, and without Jason or Steve's testimony, she had to rely on old-fashioned police interview techniques to get what she wanted. But ideally, she needed Michelle apprehended first.

There was a buzz of positivity in the room as they saw the end in sight. They were no longer scrabbling around in the mire of overwhelming and disparate pieces of evidence which didn't fit together, they now had their prime suspects. Kelly threw out orders and gave out jobs, taken gladly by the dozen uniformed officers who'd been scheduled to work the weekend, not that they expected the boss to be in. Unsocial hours certainly went by quicker when there was something to do.

Rob's mic went live and burst into life. Kelly watched as the footage from Rob's jacket was fed back to the screen. She spotted Johnny tightening harnesses and using a map to point access routes out. She felt confident knowing he was in charge. There were only so many directions Michelle could have taken.

Information came in to the incident room from several sources, and Kelly demanded that all of them were directed through her so she could update the investigation in real time, and brief all those present to keep them up to speed. The dozen or so pairs of eyes and ears in the room stood a better chance of working effectively if they worked together.

Kate approached her.

'Kelly,' she said.

Kelly knew it must be important if Kate was using her first name.

'We've had this from the search of Steve Thompson's house.'

'What is it?' Kelly asked.

'It's a property portfolio. He must have been minted.'

Kelly looked at the file and flicked through the dozens of properties owned by Steve Thompson, wondering where the hell he got all his money from. One name refused to leave her brain: Dave Crawley.

'Look at this,' Kate said, pointing at particular pages. Then Kelly saw what Kate had been trying to tell her. Parkie's was on the list, and so was Paul Gordon's house in Pooley Bridge.

Chapter 59

Johnny took the lead and headed up the Whit Beck, followed by three others including Rob Shawcross. They knew that the woman they were looking for would still be on foot because Kelly had sealed the surrounding roads off with squad cars, and there was no way the woman could have doubled back in that time. From the lake at Parkie's the terrain to the north was hostile, and Kelly had informed them that the woman was wearing only a tracksuit. Michelle had no walking kit and so would get desperate pretty quickly. She'd especially need water in the heat, and so Johnny figured that she'd stick to the rivers and becks with this in mind. However, with the roads blocked, and without going over the peaks of Lonscale and Skiddaw, which would be suicide in the heat, she had few options, and Johnny was optimistic they'd apprehend her soon enough. Of course, his brief wasn't to get involved in that side of things – that's why the police were there. It was simply his job to guide and assist.

Rob was a fit guy but he looked tired to Johnny. Kelly had mentioned that he had a habit of sleeping at the office, and now he'd been kicked out of his house after a long period of strain in his marriage. If Johnny had his way, he would have selected somebody else to accompany him, but it wasn't his job to tell the police what to do and how to do it.

'This way,' Johnny said.

They heard the helicopter overhead and it flew low and in circles, using its heat-seeking technology to determine where a human body might be hiding. Unless the woman found a rock to crawl under, the thermal imaging would spot her soon. There was no way she could have gone far on foot, especially with the ascents around here being some of the steepest in the national park.

Johnny thought about what he would do in the same position. If it were him on the run, he'd avoid elevation and stick to the cover of trees, and move slowly, following water. The woman had grown up here and so she knew the land, but Kelly had blocked the A591 to the north, and there were no other roads within reach inside a circumference estimated by the team, roughly calculating how far a woman of her stature and estimated fitness could get on the time span they were working on, as the clock ticked.

The police liaised through their radios and Johnny listened in.

A positive heat source came through from the helicopter, and coordinates were given as halfway up Blease Fell. Given it was a Saturday in mid-summer, they had to be prepared for walkers and climbers to hamper their search. It could be like trying to find one ant among thousands in the nest, but they were determined to check out each sighting. They watched above as the helicopter circled and the news came back that it was a false call.

However, five minutes later another call came through, and this time it was a sighting of a woman on her own, moving slowly, and she appeared to have no proper kit.

Johnny was familiar with the location and directed his small group east over a col, which was less than a thousand feet. As they rounded Lonscale Fell, they saw the helicopter hovering and Johnny led his group across steady terrain, around the heads of two more becks. The beds were dry, thanks to the months without rain, and the ground sure underfoot. He glanced behind as the others caught up and stayed on pace. A group of walkers stood stock still as they passed, and Rob advised them to descend the way they'd come, avoiding the foothills of Blencathra, which loomed in the distance.

There were three craggy cliffs leading of the peak, and the helicopter reported that the woman was clambering across rocks on the one closest to them. Johnny knew that she was in danger, and he ran ahead. Rob followed closely behind.

Johnny knew from experience that emotional hiking was a bad idea. It felt good, at first, to beat out one's troubles with the stomping of the feet, each step exorcising one's fears and nightmares, but that was exactly why it was dangerous. The rocks and sheer drops above Blease Gill were notorious for accidents because people weren't prepared, or they were complacent. A woman on the run would no doubt be both, as well as terrified of getting caught.

Johnny felt his lungs screaming and his thighs burning, but he spotted her. The woman was about a hundred feet away from him and had no climbing kit, and no bag. She spotted him, and the uniformed officers behind him, and she panicked.

'Stop! For God's sake!' Johnny shouted, but the breeze misdirected his cry and she didn't hear him.

He cupped his hands and shouted louder. 'Wait!' he hollered. She heard him this time and stopped. He could see her chest heaving up and down with the exertion, and he thought she surely must know the game was up. But he also knew that somebody with nothing to lose could easily choose to keep walking straight off the precipice they'd created for themselves.

'Michelle!' Rob shouted from behind him.

They watched as the woman sat down on a rock.

Johnny put his hands up and walked towards her. Rob followed.

'Mate, I'll go over there alone, I know it, it's cheeky,' he said to Rob.

'I'm good, I know these hills too, I'll follow. I need to make the arrest.'

Chapter 60

Kelly watched Rob's live feed from the viewpoint of his chest. She saw Johnny tell him to stay back and she radioed across the channel, ordering that Rob do just that. But he ignored her. She watched as the screen filled with the bobbing and blurring of a camera that was being knocked from side to side, as the wearer ran and jumped from rock to rock. She watched as Johnny crossed boulders the size of her car, and she shouted through the radio for both of them to hold back.

With eyes on Michelle, they needed a negotiator up there to talk to her. Two blokes charging at her wouldn't bring her down, and the chances of one of them having an accident, or Michelle attacking them, were extremely high. Too high to risk. Like Steve Thompson, Michelle knew her game had come to an end.

An extensive pattern of calls had emerged between the two on Tuesday evening, and Kelly wondered if Jason had already been dead by then. The coolness she'd observed in Michelle on Wednesday morning, she now knew, was all an act, probably agreed with Steve Thompson, until the noose got tighter and Steve could no longer cope with his guilt. Throw drugs and alcohol into the mix, as well as his well-documented mental health issues, and Steve thought he was better off dead than publicly guilty. This way, the debate could rage and he would never be formally charged. Perhaps Michelle was now feeling the same way.

She watched as Michelle remained seated on the lone rock. She seemed to be sobbing. Then she put her hands up and began to descend to where Johnny and Rob were. Kelly could see walkers watching from one of the cols high up on Blencathra. To them it must look like a rescue; they'd learn otherwise on the news tonight.

'Let her come to you,' Kelly instructed Rob.

She watched as Johnny put his hand behind him to stop Rob from overtaking him. She knew him so well that Kelly appreciated how frustrated he'd be right now, not being able to command the situation. It was volatile, and she told Rob to get the rescue team behind him, so he could make the arrest once, and only if, it was a genuine and safe surrender. Meanwhile the helicopter winched down a harness, as they had been instructed to, in case the woman was injured at all. Kelly heard more voices on Rob's audio and was satisfied that he wasn't making decisions alone. In these situations, calm control was vital. Michelle had murdered a man with a spade, they had no idea what else she was capable of.

They were perhaps ten feet away from her now and Kelly could make out Michelle's face. It was red with heat and exertion, and her clothes were filthy from the terrain she'd run and waded across. Her hair was dishevelled and she looked like she was exhausted and somewhat confused. She looked spent and ready for it to all be over.

Kelly instructed Rob to go easy. They had her. It was over.

'Michelle, we need you to come down safely, then we'll get you some attention,' Rob shouted.

Kelly watched Michelle respond and nod. She stood opposite a drop, and Kelly saw Johnny reach his hand to her. She held her breath, knowing that he couldn't help himself, it was his job after all.

Michelle put her foot across a gap in the rock, and the wind blew her hooded top, which she'd tied around her waist, so the zip slapped her leg each time it flapped. Johnny took one hand, and Rob stretched his hand out on the other side. She took it and Kelly and her team breathed a sigh of relief. There was no talk of arrest or charges, or what she'd done, just soothing, reassuring gentleness from human beings trying to get a woman out of a terribly dangerous situation. There was nowhere for her to go now. Kelly could see by her demeanour that all Michelle wanted was to be taken off the mountain. It was the most honest thing she'd seen in the woman's face all week.

Then she slipped, and Rob's camera went careering off the rock and crashed with a thud. They heard Michelle scream, and the sound of a man's voice trying to speak but not being able to because the wind was knocked out of him, and the brutal crashing of flesh onto rock.

The camera picked up their fall, and all Kelly could see was boulder and earth appearing on the camera, speeding past as they tumbled. In the office, none of them knew who else, apart from Michelle, had fallen.

Chapter 61

Michelle Parkinson's first police interview was at her hospital bedside, a week later.

She'd sustained a broken arm, fractured ribs, a collapsed lung and a broken foot, but she was alive. She'd survived because Rob had taken the brunt of the fall. His body had protected hers, and had provided a cushion for her as they careered down the rocky cliff. Johnny said later that all he could do was watch as they both slipped from his grasp.

Rob had been taken to the Penrith and Lakes in the helicopter that had first found Michelle Parkinson hiding on the mountainside. He'd been the priority casualty. He'd survived for fourteen hours. His wife had been by his side as he passed away.

Kelly had wandered through the next few days on autopilot, not wanting to face the awful reality of the loss of one of her team, not just that but an inspiring human being. Rob had been kind, funny, hard-working and honest. Eden House remained in shock, in mourning for a standout officer, and one who, in the end, had put his job before himself. Kelly blamed herself for allowing him to be there, knowing he was exhausted, and his mind not properly on the job. Everybody from Kate to the superintendent and Johnny had assured her that it wasn't her fault. But the burden of guilt weighed heavy and she found it increasingly difficult to get out of bed.

Now, as she sat opposite Michelle Parkinson's hospital bed, the rage inside her threatened to get the better of her. The one thing stopping it winning, and turning her into the same sort of monster she caught and put behind bars, was the fact that she knew that's what had happened to Michelle, and that's why Brian and Jason were dead.

Michelle was a woman wronged, there was no doubt about that. The abuse from her father, followed by her mother giving up, and her growing mistrust of those who were in a position of authority and meant to protect her, like Steve Thompson, all conspired to make a monster that used hate as a tool of revenge. Kelly wasn't about to become the same. It would have been so easy to take a chair and slam it across Michelle's broken bones, just to rid herself of the fury that boiled within her body.

She was different. And Rob's memory deserved more.

Michelle opened her eyes and looked startled to see Kelly, and also fearful, as if expecting some kind of punishment there and then, and for it to be painful. Kelly smiled.

'Michelle.'

'Kelly.' Her eyes darted around, settling on the restraints on her bed.

'Don't worry, I'm not going to hurt you, though you clearly think I should.'

Before Michelle had stirred, Kelly had moved the nurse's alarm beyond her reach. Michelle noticed this now.

'I'm not going to hurt you, but you are going to tell me everything, and I'll leave when you've done that. You ranted and mumbled as you were brought in, but it wasn't recorded. I'm here to record your statement,' Kelly said, taking recording equipment out of the bag she'd brought with her. She felt empty inside, but covered in steel armour, which held her up, at least long enough to get this woman to confess.

'So, when you're ready,' Kelly said.

Kelly had gone to the hospital with little confidence that Michelle would talk. She hadn't planned on getting much. She hadn't even considered if it was a waste of time taking the recording equipment, she'd just acted on instinct. All trepidation about what she might or might not get from this witness had gone with Rob's last breath. Her pulse was below seventy, and her face unflinching as Michelle began to talk.

Michelle described how she'd pleaded with Brian at Calfclose Bay to drop his stupid pursuit of Thompson. She'd warned him, but Paul had called Jason and told him what they'd talked about. Jason was waiting for Brian when Michelle delivered him to Jason's car, and Thompson was ready to take him for a drive. Michelle maintained she wasn't with them when they drove to Thirlmere. Steve Thompson had been an associate consultant on the draining of the reservoir, and he knew just the right spot, where the earth would be soft enough to dig, before it was covered over again by millions of gallons of water.

Brian was drunk so he didn't resist, or so Jason told her. It was Steve's idea to cut his leather jacket because Jason, being such a fuckwit, wanted to keep it for himself, having always coveted it. Steve cut it up so he couldn't. They bound Brian with the cut-offs, and Thompson was the one who delivered the blow from the same spade they used to dig the grave. Jason allegedly laughed the whole way through it, having been given speed by Thompson, whose supplier was a guy who worked for the Crawley family. Dave already, at seventeen, was becoming a mule and making a shit-ton of money from it.

Michelle had kept her secret for twenty-four years, but she and Jason had sunk into a pit of drug-induced despair, so neither had to face their demons. Thompson bought Michelle a place, as well as Paul Gordon, and they all swore secrecy.

It worked, until drought played its hand. Jason was the first to panic. And Michelle snapped. Kelly got her on record admitting to his murder. She swore she didn't go near Steve Thompson all week, but she knew who still supplied him with his drugs.

Paul Gordon wasn't a plasterer by trade, that was for sure. During a raid on his home, police seized over 3,000 synthetic opioid pills. The vast majority of them were fentanyl, which mimicked the effects of heroin and opium when crushed, snorted, injected or simply swallowed, but could be fifty to a hundred times more potent than morphine and thus often lethal. Toxicology results on Steve Thompson revealed the same compound in his blood and had been ruled his cause of death. His blood alcohol was also seven times over the legal driving limit. The combination would have killed an ox. Steve Thompson had been a dead man walking anyway, as his liver was in the advanced stages of primary biliary cirrhosis. Ted reckoned he probably had a year until total failure. A lifetime of guilt and shame, swallowed and injected, until he felt nothing. Same for Jason Cooper. What they did to Brian twenty-four years ago haunted them and chased them down a hole of self-destruction, from which they knew they could never climb out.

'Did it feel good?' Kelly asked Michelle.

'What?'

'Killing Jason? Hitting him over the head as hard as you figured he'd done to Brian?'

'I told you Steve did that,' Michelle said.

'But you never knew for sure, did you? Who actually delivered the blow? It could have been Steve, because he was sober and strong, and focused enough to get it done, but for you, it might as well have been Jason.'

Michelle nodded.

Kelly stood up.

'When you're well enough to be discharged, you'll be charged with murder and processed at a convenient station, then you'll have a hearing before a judge, where you'll enter a plea, then you'll be kept in custody, a nice female prison if you're lucky, until your trial date. Goodbye, Michelle.'

Kelly turned away.

'I'm sorry about your friend,' Michelle said. Kelly stopped, but she didn't look back. The thing with compassion was that you had to give it to everybody, but this time Kelly couldn't find it in herself. The capacity for duality, when it suited the criminal, staggered Kelly and she walked out of the room, thanking the guard outside.

Instead of forgiving Michelle, and perhaps smiling or nodding to accept and dissipate her burden, Kelly kicked a gurney in the corridor on her way out, and it hurt like hell.

Chapter 62

'Hi Finch, it's good to see you. You look well. You remember Ted Wallis?'

Kelly had arranged the meeting in The Crown, in Pooley Bridge. Mick Finch had worked in the serious crime unit as a contemporary with her father. Ted remembered the young uniform standing in on the odd autopsy. The coroner had an excellent memory for faces, and it was probably something to do with the way he got to know them intimately when they stared at him from his slab. Kelly recalled Finch only as a copper who used to call round for John Porter, before they went for a pint together. He was in his seventies now and happily retired.

She'd spent the morning at Rob's funeral.

The occasion had left her empty and numb. It was unfortunate that she'd scheduled the meeting with Finch for the same day, but the funeral date was only released last week. Kelly had begged her father not to autopsy him. She couldn't bear the thought of him being cut up and poked about, even with respect, for what was ultimately a foregone conclusion. He agreed and signed off Rob's death certificate.

The whole constabulary had turned out. Andrew Harris had given a fitting address to the press.

But they felt they'd never get over it. The office would never be the same. No one wanted to sit in his chair, or turn on his computer, but they'd had to, to finalise the work he'd done on the murders. Kelly always knew that Rob crunched numbers and data like her daughter went through rusks, but it still astounded her the sheer volume of information Rob was able to process at any given moment in an investigation. She was the one who took his seat and trawled through his work.

The echoes of banter between Rob and Dan sat on the desktops like dust, and it became very quiet. Kelly had seen Dan push back on his wheelie chair more times than she could count to visit Rob's desk, only to find him not there. Uniforms, drafted in for the two cases, walked around Rob's desk respectfully and in silence as they passed.

It was a fatigue that was unmatched by anything else she'd ever encountered. She felt as though somebody had washed her out and wrung her dry. Given the choice, she'd like to curl up and fall into a deep sleep, on the floor of The Crown, if necessary.

But she knew this meeting was important, because Mick Finch was due to give evidence at the internal affairs tribunal being held on the conduct of John Porter with reference to historic cases, in three days.

Finch had already signed his statement. This was an off the record gathering. Kelly guessed it was his way of trying to ease the burden and explain some of the things he was ashamed of. Finch wasn't implicated in the misconduct hearing. He was solely a witness. Like most internal investigations inside enormous institutions, they only needed one fall guy, and John Porter was their guy. He'd been in charge – though he shouldn't have been – and the book stopped there.

It had been taken out of Kelly's hands, not only due to conflict of interest, but also because the case had been passed to fraud. It was standard. She knew there was no way she'd be kept on the investigation, and nor did she want to be. But it was strangely comforting to see Finch, who she could place in her past and tell herself that it wasn't all bad.

They sipped their drinks and Finch told stories about the good old days, though they were tinged with an edge of wariness. He tried to keep to the ones that were benign, and proven to be so. Kelly could tell that he was trying to make her feel better. Ted recognised some of the stories.

'That reminds me,' Finch said.

Kelly waited.

'I got wind that it was a stolen locket that finally brought Michelle Parkinson down. I did a bit of digging – I hope you don't mind Kelly – and I've seen the photograph.'

Kelly did mind. It reminded her that old timers still had access to confidential information, and this was exactly the kind of practice they were trying to stop. However, she was still keen to hear what he had to say.

'It wasn't stolen,' Kelly said. 'It was her mother's. An heirloom.'

'It most certainly was not. It was part of a heist taken from a pad on Derwent back in 1978. I was on the case. The locket was on the inventory. Items like that stick in your head.'

Kelly's gut churned. She knew what was coming. 'You're going to tell me that it was an original, aren't you?' she asked.

Finch nodded. 'It's probably worth about half a million now.'

'Our expert thought about a hundred thousand, until we couriered it to him, then he said exactly what you're telling me now. I knew Ken Parkinson couldn't have afforded that thing, which is why I believed Michelle when she told me she'd got it from her mother's mother.'

Finch shook his head.

Kelly downed her drink and went to the bar to order another. She felt Ted's eyes on her back. Not only was John Porter implicated in abuse of process and misconduct, now she knew he'd been on the take too, gifting the locket to his mate, who gave it to his wife, Michelle's mother. She didn't need to ask Finch for evidence. The board would find it eventually.

Kelly walked back to her seat, and Ted put his hand lovingly over hers. They couldn't help but stare out to the River Eamont, which went benignly on its merry way, no matter the season. Behind that, the forest beckoned to Kelly. They sipped their drinks and she heard Ted telling Finch about his friend Henry who'd been invaluable in their inquiries. Henry had stuck around for a while, and they were planning a road trip together, for Ted to show him some of the Lakes' highlights. She felt hot salty tears well up in her eyes, but she forced them back. She got up and Ted squeezed her hand.

'I think I might need some time off,' she said.

'I think that's a grand idea,' Ted said.

Acknowledgements

A few long hot summers inspired the story which led to *Silent Bones*, that and a beautiful walk up Raven Crag. The Lake District never fails to offer new ideas.

I'd like to thank Adrian Priestley for being on call at all times to answer my questions.

Thanks to the fantastic team at Canelo, especially Alicia Pountney for all the hard work that goes into the editing process, as well as Tom Sanderson for the stunning artwork on the Kelly Porter series.

Thank you to my agent, Peter Buckman, for his tireless loyalty to me and Kelly. Also to my biggest fans, Mike, Tilly, Freddie and Poppy, and all the people around me who put up with endless discussions about crime and policing.

Do you love crime fiction and are always on the lookout for brilliant authors?

Canelo Crime is home to some of the most exciting novels around. Thousands of readers are already enjoying our compulsive stories. Are you ready to find your new favourite writer?

Find out more and sign up to our newsletter at canelocrime.com